Key Questions in Mammalogy:
A Study and Revision Guide

Key Questions in Mammalogy: A Study and Revision Guide

Paul A. Rees

CABI is a trading name of CAB International

CABI	CABI
Nosworthy Way	200 Portland Street
Wallingford	Boston
Oxfordshire OX10 8DE	MA 02114
UK	USA
Tel: +44 (0)1491 832111	Tel: +1 (617)682-9015
E-mail: info@cabi.org	E-mail: cabi-nao@cabi.org
Website: www.cabi.org	

The views expressed in this publication are those of the author(s) and do not necessarily represent those of, and should not be attributed to, CAB International (CABI). Any images, figures and tables not otherwise attributed are the author(s)' own. References to internet websites (URLs) were accurate at the time of writing.

CAB International and, where different, the copyright owner shall not be liable for technical or other errors or omissions contained herein. The information is supplied without obligation and on the understanding that any person who acts upon it, or otherwise changes their position in reliance thereon, does so entirely at their own risk. Information supplied is neither intended nor implied to be a substitute for professional advice. The reader/user accepts all risks and responsibility for losses, damages, costs and other consequences resulting directly or indirectly from using this information.

CABI's Terms and Conditions, including its full disclaimer, may be found at https://www.cabidigitallibrary.org/terms-and-conditions.

A catalogue record for this book is available from the British Library, London, UK.

ISBN-13: 9781800628656 (hardback)
       9781800624528 (ePDF)
       9781800624535 (ePub)

DOI: 10.1079/9781800624535.0000

Commissioning Editor: Ward Cooper
Editorial Assistant: Emma McCann
Production Editor: James Bishop

Typeset by Straive, Pondicherry, India
Printed and bound in the UK by CPI Group (UK) Ltd, Croydon, CR0 4YY

For Elliot –
the budding zoologist (and football star)

Contents

About the author

Paul Rees was a senior lecturer in the School of Science, Engineering and Environment at the University of Salford, United Kingdom, for 22 years until his retirement in 2020. He holds a BSc in Environmental Biology from the University of Liverpool, a PhD in animal ecology and behaviour from the University of Bradford and an LLM in Environmental Law from the University of Central Lancashire. Paul previously lectured at three Further Education Colleges and a Higher Education College in the United Kingdom, and trained biology teachers at Sokoto College of Education in Nigeria. He has taught from GCE 'O'/GCSE level to MSc level and has been an external examiner for a range of taught programmes, from Higher National Diploma to MSc level, at six British universities. Paul has published papers on mammal behaviour and ecology (especially felids and elephants), wildlife law, and the role of zoos in conservation, along with fifteen textbooks concerned with ecology, zoo biology, wildlife law and elephants. He is the author of CABI's *Students' Dictionary of Zoo and Aquarium Studies* and also six other titles in CABI's *Key Questions* series:

Key Questions in Ecology
Key Questions in Applied Ecology and Conservation
Key Questions in Biodiversity
Key Questions in Zoo and Aquarium Studies
Key Questions in Animal Behaviour and Welfare
Key Questions in Wildlife and Nature Conservation Law

Preface

Mammalogy is an important component of many bioscience programmes at undergraduate and postgraduate levels. Many programmes of study include a consideration of mammals either as a discrete module or as part of modules in vertebrate zoology, ecology, behaviour and related subjects.

Large mammals are ecologically important because many are keystone species and ecosystem engineers and, as such, they have a disproportionate effect on the structure, composition and stability of natural habitats. They are apex predators in many terrestrial, freshwater and marine ecosystems, and they provide us with food, companions and research animals. Mammals are the most popular animals kept in zoos and are the focus of much of our concern about the loss of species diversity.

As mammals ourselves, we are interested in the origins, anatomy, evolution and genetic relationships within the mammals, especially apes, and we share some of their parasites and diseases.

This book presents 600 multiple-choice questions on mammalogy. It begins by considering the biological characteristics of mammals and the history of mammalogy as a science. It continues by examining the origin and evolution of mammals and their taxonomy, followed by aspects of their anatomy and physiology. Mammal behaviour, ecology and genetics are considered next and then their zoogeography. The conservation and management of wild mammal populations are then examined, followed by a consideration of their parasites and diseases and finally their relationship with humans through domestication and our exploitation of various species as food, companion animals and for other reasons.

This book is not intended as a series of tests. Inevitably the details taught on a particular programme of study will vary depending upon the nature of the course and the details of the syllabus. The purpose of the book is to make the reader aware of certain facts, processes and principles in mammalogy and I hope I have included a sufficiently varied range of questions to make the book of use to anyone studying this most important group of animals.

Acknowledgements

I am extremely grateful to Ward Cooper (Commissioning Editor at CABI) for commissioning this book and for his support, and that of his colleagues Emma McCann (Editorial Assistant) and James Bishop (Senior Production Editor), during its production.

Most of the images used in this book are my own. Almost all of the images of mammal skulls, teeth and bones in Chapter 3 are of specimens held by Bolton Museum and I am grateful to Ian Trumble for allowing me access to them. Fig. 7.9 is part of an image of Alfred Russel Wallace which is in the public domain and was first published in *Borderland Magazine* in April 1896. The images used in Fig. 2.9 (Tasmanian devil) and Fig. 7.5 (fossa) were taken by my daughter Clara and I am most grateful for her permission to use them.

Finally, I should like to thank my grandson, Elliot, who helped me choose the cover image and for giving me an excuse to visit zoos far more often than I could otherwise have justified.

How to use this book

The questions in each chapter are divided into three levels: foundation, intermediate, and advanced. These levels are not intended to reflect any particular curriculum but rather general levels of difficulty, and should not be taken too seriously. Knowledge of basic facts are dealt with at the foundation level while the intermediate level and advanced levels contain questions involving more obscure facts and concepts. However, there is some variation between chapters as not all of the areas covered lend themselves to this approach. Students are advised to check the syllabuses they are following in detail before relying too much on this book as a preparation for specific exams.

Students are encouraged to complete a whole chapter – or at least a complete section (foundation, intermediate or advanced) – before looking at the answers. This is because the explanations for some answers may assist in selecting the correct answer to subsequent questions, although I have tried to avoid this where possible. The order in which the chapters are attempted does not really matter because each is about a distinct topic. However, within any chapter you are advised to attempt the foundation questions first, followed by the intermediate questions and finally the advanced questions.

1 The History and Principles of Mammalogy

This chapter contains questions concerned with the origins of mammalogy as a distinct discipline within zoology, important milestones in its development and the defining characteristics of mammals.

Foundation

1.1f Which of the following is not an alternative term for mammalogy?

    a. Mastology

    b. Theriology

    c. Gerontology

    d. Therology

1.2f Which major two-volume text on mammals was first published in 1964?

    a. *Walker's Mammals of the World*

    b. *Walton's Mammals of the World*

    c. *Whitaker's Mammals of the World*

    d. *Warwick's Mammals of the World*

1.3f Which of the following pairs of terms describes the physiology of mammals?

    a. Endotherm and ectotherm

    b. Homiotherm and poikilotherm

© Paul A. Rees 2024. *Key Questions in Mammalogy* (P.A. Rees)
DOI: 10.1079/9781800624535.0001

   c. Endotherm and homiotherm

   d. Ectotherm and poikilotherm

1.4f **Which of the following are not normally characteristics of mammals?**

   i. A diaphragm

   ii. Two bones in the middle ear

   iii. Hair

   iv. Four-chambered heart

   v. Mammary glands

   vi. A right aortic arch

   a. i, ii and iii only

   b. vi only

   c. ii and vi only

   d. ii, iv and vi only

1.5f **The circulatory system of a mammal is**

   a. open and single

   b. open and double

   c. closed and single

   d. closed and double

1.6f ***The Naked Ape* is a book by**

   a. Desmond Morris about humans

   b. Jane Goodall about chimpanzees

   c. Dian Fossey about gorillas

   d. Frans de Waal about orangutans

1.7f **The 'National Mammal' of the United States is the**

   a. black bear (*Ursus americanus*)

   b. American bison (*Bison bison*)

    c. puma (*Puma concolor*)

    d. grey wolf (*Canis lupus*)

1.8f   **The term 'mammal' is derived from the Latin for**

    a. warm-blooded

    b. hair

    c. milk

    d. breast

1.9f   **According to Schmidly and Naples (2019) the early North American mammalogists were mostly trained as**

    a. teachers or engineers

    b. physicians, artists or clergymen

    c. scientists or teachers

    d. soldiers and explorers

1.10f **Which of the following disciplines are not legitimate areas of interest for mammalogists?**

    i. systematics

    ii. palaeontology

    iii. behaviour

    iv. physiology

    v. ecology

    vi. anatomy

    vii. biochemistry

    a. ii and vii

    b. ii, iii and v

    c. i, iv, vi

    d. All of these disciplines are of interest to mammalogists

1.11f Who believed that giraffes evolved long necks because successive generations reached progressively higher in the trees to reach leaves (Fig. 1.1)?

Fig. 1.1.

    a. Charles Darwin

    b. Alfred Russel Wallace

    c. Jean-Baptiste Lamarck

    d. Thomas Huxley

1.12f **The smallest extant mammal species in the world by mass is a species of**

    a. bat

    b. shrew

    c. mouse

    d. possum

1.13f **According to the Mammal Society (2024), how many mammal species are found in and around the British Isles?**

    a. 53

    b. 79

    c. 107

    d. 123

1.14f **The term 'Mammalia' was first used by**

    a. Carl Linnaeus

    b. Aristotle

    c. Jean-Baptiste Lamarck

    d. Richard Owen

1.15f **Which of the following structures is unique to mammals?**

    a. Diaphragm

    b. Pancreas

    c. Spleen

    d. Caecum

1.16f **In 1986 the British zoologist Ernest Neal published *The Natural History of***

    a. *Foxes*

    b. *Red Deer*

    c. *Brown Hares*

    d. *Badgers*

1.17f Baird's tapir (*Tapirus bairdii*) is named after Spencer Fullerton Baird. He was

    a. a Scottish zoologist who worked at the British Museum (Natural History)

    b. an American naturalist who worked at the Smithsonian Institution

    c. an English explorer who collected animals for the Zoological Society of London

    d. a Welsh zoologist who worked at the University of Manchester

1.18f Who has been described as the 'pioneer naturalist of Australia' and published *Mammals of Australia* in 1863?

    a. Elizabeth Coxen

    b. Morton Allport

    c. John Gould

    d. Charles Darwin

1.19f The British Library holds drawings of mammals and other species collected in Sumatra by

    a. Charles Darwin

    b. Thomas Stamford Raffles

    c. Richard Owen

    d. Thomas Huxley

1.20f *The History of British Mammals* was written by

    a. David Macdonald

    b. H. N. Southern

    c. Derek Yalden

    d. L. Harrison Matthews

Intermediate

1.1i   In 1901 the British explorer Sir Harry Hamilton Johnston sent the first pieces of hide of which of the following species (Fig. 1.2) to the British Museum?

Fig. 1.2.

     a. A

     b. B

     c. C

     d. D

1.2i Which of the following zoologists conducted a detailed study of the ecology and behaviour of lions (*Panthera leo*) in the Serengeti?

    a. George Schaller

    b. Bernhard Grzimek

    c. Iain Douglas-Hamilton

    d. Keith Eltringham

1.3i The scientific journal published by the Mammal Society is called

    a. *Mammal Biology*

    b. *Mammal Studies*

    c. *Mammal Review*

    d. *Mammal Science*

1.4i Match the species in Table 1.1 with the date each was discovered by scientists.

    a. A

    b. B

    c. C

    d. D

Table 1.1

Species	A	B	C	D
Vu Quang ox	2003	1869	1965	1992
Giant panda	1965	2003	1992	1869
Kipunji	1992	1992	1869	2003
Iriomote cat	1869	1965	2003	1965

1.5i Mammalian hair consists mostly of

    a. keratin

    b. collagen

    c. myosin

    d. fibrinogen

1.6i Temminck's flying fox (*Pteropus temminckii*), Temminck's pangolin (*Smutsia temminckii*) and Temminck's golden cat (*Catopuma temminckii*) are all named for C. J. Temminck, a

    a. British naval officer

    b. Dutch zoologist

    c. German colonial administrator

    d. Belgian explorer

1.7i The first person to coin the term 'primates' was

    a. Carl Linnaeus

    b. Richard Owen

    c. Charles Darwin

    d. Aristotle

1.8i Prof. Tim Clutton-Brock is famous for his long-term studies of

    a. blue wildebeest, black bears and koalas

    b. orangutans, fallow deer and tigers

    c. red deer, meerkats and Soay sheep

    d. African lions, giraffes and chimpanzees

1.9i Mammals (and birds) in cold regions of the Earth tend to be bulkier than individuals of the same species in warm regions, because large individuals lose heat slower than small individuals. This principle is known as

    a. Müller's rule

    b. Meyer's rule

    c. Schreiber's rule

    d. Bergmann's rule

1.10i Frans de Waal published a book entitled

    a. *Chimpanzee Politics*

    b. *Gorilla Government*

    c. *Monkey Management*

    d. *Ape Administration*

1.11i Who was the author of *The Life of Mammals*, published in 1957?

    a. G. G. Simpson

    b. D. F. Attenborough

    c. D. J. Morris

    d. J. Z. Young

1.12i During which decade were major books on the ecology and behaviour of chimpanzees, gorillas, African elephants, grey wolves and African lions describing field studies first published?

    a. 1950s

    b. 1960s

    c. 1970s

    d. 1980s

1.13i *The Viviparous Quadrupeds of North America (1846–1854)* was the most significant publication on North American mammals available in its day and was produced by

    a. John Bachman and John James Audubon

    b. Ernest Walker and Harrison Allen

    c. John Godman and James De Kay

    d. Richard Harlan and Joseph Grinnell

1.14i The Smithsonian Institution was founded in the United States in 1846. Important collections of mammal specimens were deposited in the Smithsonian in the 1850s by members of the US Army Medical Corps who accompanied

    a. prospectors migrating to California seeking gold deposits

    b. cattlemen moving livestock to the Midwest

    c. surveying parties seeking routes for a transcontinental railroad

    d. naturalists working for scientific societies

1.15i **Dr John Napier (1917–1987) was a leading British authority on**

a. marsupials

b. equids

c. felids

d. primates

1.16i **Raymond Dart, Mark Leakey and Eugène Dubois were all famous for their work on the fossil history of**

a. elephants

b. humans

c. marsupials

d. ungulates

1.17i **Which of the following is a new species of bovid whose remains were found in 1992 prior to a live specimen being recorded in 2010?**

a. Lesula

b. Saola

c. Kipunji

d. Khanyou

1.18i **The scientific name *Puma concolor* was proposed for the cougar by the Scottish naturalist William Jardine in 1834 because of the absence of any living near relatives. The original name assigned to this species by Linnaeus was**

a. *Panthera concolor*

b. *Profelis concolor*

c. *Felis concolor*

d. *Acinonyx concolor*

1.19i **Which American zoologist is considered by many scholars to have been the 'academic father of mammalogy'?**

a. Joseph Grinnell

b. Maureen Downey

c. Alberta Seaton

d. Elliott Coues

1.20i **The wing of a bat, pectoral fin of a dugong and the forelimb of a mole are homologous structures built on the same basic plan. In 1843, who defined homologues in zoology as 'the same organ in different animals under every variety of form and function'?**

a. Richard Owen

b. Charles Darwin

c. Mary Anning

d. Alfred Russel Wallace

Advanced

1.1a **Eadweard Muybridge conducted pioneering work on the gait of mammals in the last quarter of the nineteenth century (Fig. 1.3). He was**

a. an anatomist

b. a zoologist

c. a physicist

d. a photographer

Fig. 1.3.

1.2a **Which American zoologist and palaeontologist published** *The Principles of Classification and a Classification of Mammals* **in 1945?**

    a. Elizabeth Adams

    b. George Gaylord Simpson

    c. Olive Griffith Stull

    d. Benjamin Preston Clark

1.3a **The official journal of the American Society of Mammalogists is the**

    a. *Mammalian Biology*

    b. *Mammal Notes*

    c. *Journal of Mammalogy*

    d. *Mammal Research*

1.4a **Which of the following species was 'discovered' or identified as a new species by western explorers or naturalists first?**

    a. Bonobo (*Pan paniscus*)

    b. Pygmy hippopotamus (*Choeropsis liberiensis*)

    c. Okapi (*Okapia johnstoni*)

    d. Giant panda (*Ailuropoda melanoleuca*)

1.5a **In the early twentieth century a clash developed over research and scientific ethics between factions within the American Society of Mammalogists over**

    a. the predator control practices of the Biological Survey

    b. the specimen collecting policy of the Smithsonian

    c. the focus of American universities on the study of large mammals

    d. the lack of serious interest in conservation exhibited by some members

1.6a **Who published *The Age of Mammals in Asia, Europe and North America* in 1910?**

a. Henry Fairfield Osborn

b. Roy Chapman Andrews

c. George Gaylord Simpson

d. Lee R. Dice

1.7a **Which of the following was not a well-known American mammalogist?**

a. Joseph Grinnell

b. Vernon Orlando Bailey

c. Leonard Harrison Matthews

d. Edward William Nelson

1.8a **The fossil of Piltdown Man that was 'found' by the amateur archaeologist Charles Dawson in 1912 was thought to have been the 'missing link' between early apes and humans but was a fake consisting of**

a. the cranium of a young gorilla and a human jaw and teeth

b. the mandible and teeth of a chimpanzee and the maxilla of a human

c. the teeth of a gorilla and the maxilla of an orangutan

d. the mandible and teeth of an orangutan and a human cranium

1.9a **Match the book titles below with their authors (Table 1.2)**

Table 1.2

Title	A	B	C	D
Mammal Societies	M.J Delany and D.C.D Happold	L.D. Mech	R. F. Ewer	T. Clutton-Brock
Ethology of Mammals	L.D. Mech	T. Clutton-Brock	M.J Delany and D.C.D Happold	R. F. Ewer
Ecology of African Mammals	T. Clutton-Brock	R. F. Ewer	T. Clutton-Brock	M.J Delany and D.C.D Happold
Wolves – Behavior, Ecology and Conservation	R. F. Ewer	M.J Delany and D.C.D Happold	L.D. Mech	L.D. Mech

a. A

b. B

c. C

d. D

1.10a Which of the following divided the animals we now call mammals into the viviparous quadrupeds, cetaceans and humans?

a. Galen

b. Aristotle

c. Lamarck

d. Hippocrates

1.11a The American mammalogist William Henry Burt (1903–1987) is best known for making important contributions to our understanding of the concepts of

a. interspecific competition in mammals

b. the taxonomy and distribution of mammals

c. home range and territoriality in mammals

d. life history traits and the ecology of mammals

1.12a Who published a monograph entitled *General Report on North American Mammals* in 1859 in which more than 730 species of mammals were described, many of which were discovered in railroad surveys searching for the best route westward to the Pacific Ocean?

a. Edgar Alexander Mearns

b. Spencer Fullerton Baird

c. Thomas Say

d. Major Stephen Long

1.13a Who is considered to be an expert in the evolutionary basis of female behaviour in humans and nonhuman primates?

a. Sarah Hrdy

b. Jane Goodall

    c. Dian Fossey

    d. Biruté Galdikas

1.14a **Mammals have more complex brains than other vertebrates and primates have the most complex brains of all (Fig. 1.4). What is the name of the hypothesis that predicts that species with relatively large neocortices should exhibit more complex social strategies than species with smaller neocortices?**

Fig. 1.4.

    a. Social Brain Hypothesis

    b. Social Neocortex Hypothesis

    c. Strategic Brain Hypothesis

    d. Social Mind Hypothesis

1.15a **The Societas Europea Mammalogica was founded in**

    a. 1888

    b. 1908

    c. 1958

    d. 1988

1.16a The Australian Mammal Society was founded in

> a. 1948
>
> b. 1958
>
> c. 1968
>
> d. 1978

1.17a In which year was the *Atlas of European Mammals* published?

> a. 1977
>
> b. 1988
>
> c. 1999
>
> d. 2009

1.18a Match the name of each of the mammal societies in Table 1.3 with the date it was founded.

Table 1.3

Society	A	B	C	D
German Society for Mammalian Biology	1926	1954	1983	1983
Mammal Society (of the UK)	1954	1926	1926	1992
Russian Theriological Society	1992	1983	1954	1954
Italian Mammal Society	1983	1992	1992	1926

> a. A
>
> b. B
>
> c. C
>
> d. D

1.19a The Grinnell Method is

> a. a detailed protocol for recording observations in the field using a field notebook, a field journal, a species account and a catalogue of specimens.
>
> b. a technique used in taxidermy for mounting and preserving mammals so that they adopt a naturalistic pose

    c. a method of constructing an ethogram used to describe and record the behaviour of mammals in the field

    d. a method of locating wild mammals using radio tracking equipment

1.20a Who published *Histoire Naturelle des Orang-Outangs* in 1795?

    a. Jean-Baptiste Lamarck and Pierre Belon

    b. Georges Cuvier and Etienne Geoffrey Saint-Hilaire

    c. Richard Owen and Erasmus Darwin

    d. Konrad Gessner and Leonhart Fuchs

2 Origins, Evolution and Taxonomy

This chapter contains questions concerned with the evolutionary origins of mammals and the taxonomy of present day and extinct mammalian taxa.

Foundation

2.1f   All mammals are synapsids and possess on each side of the skull

    a. no temporal opening

    b. a single temporal opening

    c. a pair of temporal openings

    d. four temporal openings

2.2f   The term synapsid means

    a. unfused jaw

    b. fused jaw

    c. unfused arch

    d. fused arch

2.3f   Early mammals arose from mammal-like reptiles in the

    a. Jurassic

    b. Triassic

    c. Cretaceous

    d. Permian

2.4f **Stem-mammals are also known as**

    a. mammal-like reptiles

    b. protomammals

    c. paramammals

    d. all of the above

2.5f ***Miohippus* was a Miocene**

    a. horse

    b. hippopotamus

    c. rhinoceros

    d. elephant

2.6f **New World kangaroo rats (*Dipodomys*) and Old World jerboas (*Allactaga*) possess elongated hind limbs and move using bipedal hopping. These characteristics have evolved independently in the two taxa and are an example of**

    a. divergent evolution

    b. monophyletic evolution

    c. convergent evolution

    d. conflux evolution

2.7f **Which of the following is not an extinct order of mammals?**

    a. Multituberculata

    b. Plesiosauria

    c. Dinnetheria

    d. Docodonta

2.8f **The species shown in Fig. 2.1 belongs to the order**

    a. Perissodactyla

    b. Artiodactyla

    c. Pholidota

    d. Tubulidentata

Fig. 2.1.

2.9f **Which of the following is the nominotypical or nominate subspecies of the tiger?**

a. *Panthera tigris altaica*

b. *Panthera tigris sumatrae*

c. *Panthera tigris tigris*

d. *Panthera tigris corbetti*

2.10f **Which of the following is not a eutherian?**

a. Etruscan shrew (*Suncus etruscus*)

b. Western pygmy possum (*Cercartetus concinnus*)

c. Common pipistrelle (*Pipistrellus pipistrellus*)

d. Wood mouse (*Apodemus sylvaticus*)

2.11f **The earliest known mammals, morganucodontids, were small,**

a. crepuscular and tarsier-like

b. nocturnal and squirrel-like

c. nocturnal and shrew-like

d. diurnal and hedgehog-like

2.12f **Fossils of *Gigantopithecus blacki* are known from southern China. It is the largest known species of**

a. monkey

b. great ape

c. bear

d. big cat

2.13f **Fig. 2.2 shows a model of a**

a. *Smilodon*

b. *Steropodon*

c. *Glaucodon*

d. *Diprotodon*

Fig. 2.2.

2.14f **Glyptodonts are an extinct clade of**

a. pangolins

b. hedgehogs

c. aardvarks

d. armadillos

2.15f *Proconsul* **first appeared in the Miocene and was an early**

a. rhinoceros

b. antelope

c. marsupial

d. ape

2.16f **The largest extant species of rodent is the**

a. North American beaver (*Castor canadensis*)

b. Crested porcupine (*Hystrix cristata*)

c. Capybara (*Hydrochoerus hydrochaeris*)

d. Patagonian mara (*Dolichotis patagonum*)

2.17f **Jonathan Kingdon is an expert in the taxonomy and evolution of the mammals of**

a. North America

b. South America

c. Africa

d. Australasia

2.18f **The scientific names** *Panthera uncia, Uncia uncia, Felis uncia* **and** *Felis irbis* **all refer to the snow leopard (Fig. 2.3) and are referred to as**

a. synonyms

b. revisions

b. antonyms

d. homonyms

Fig. 2.3.

2.19f **Put the fossils in Table 2.1 in the chronological sequence of their first appearance in the fossil record.**

    a. A

    b. B

    c. C

    d. D

Table 2.1

Chronological sequence	A	B	C	D
Most recent ←	Homo neanderthalensis	Homo neanderthalensis	Homo habilis	Australopithecus afarensis
	Homo habilis	Homo erectus	Homo neanderthalensis	Homo habilis
	Homo erectus	Homo habilis	Homo erectus	Homo neanderthalensis
Earliest	Australopithecus afarensis	Australopithecus afarensis	Australopithecus afarensis	Homo erectus

2.20f **The mammals became the dominant animal group on Earth following an asteroid strike that caused a mass extinction which destroyed three quarters of all animal life**

    a. 42 million years ago

    b. 51 million years ago

    c. 66 million years ago

    d. 73 million years ago

Intermediate

2.1i **The extinct *Megatherium americanum* was a sloth the size of**

    a. a capybara

    b. a Cape buffalo

    c. a tapir

    d. an elephant

2.2i **Miacids were primitive**

    a. insectivores

    b. antelopes

    c. cat-like carnivorans

    d. bats

2.3i **The species shown in Fig. 2.4 is a**

    a. tarpan

    b. Przewalski's horse

    c. Persian onager

    d. quagga

Fig. 2.4.

2.4i **Which of the following was not a prehistoric elephant?**

a. *Paraceratherium*

b. *Amebelodon*

c. *Gomphotherium*

d. *Deinotherium*

2.5i **Cetaceans are thought to have evolved from**

a. even-toed ungulates

b. odd-toed ungulates

c. pinnipeds

d. sirenians

2.6i **The genus *Australopithecus* is derived from a Latin word and a Greek word, respectively, meaning**

a. 'eastern' and 'monkey'

b. 'western' and 'ape'

   c. 'northern' and 'monkey'

   d. 'southern' and 'ape'

2.7i **Which of the following structures migrated to a position under the skull during the evolution of bipedalism in primates?**

   a. The zygomatic process

   b. The ethmoid bone

   c. The mental foramen

   d. The foramen magnum

2.8i **Which of the species shown in Fig. 2.5 is not a pachyderm?**

Fig. 2.5.

a. A

b. B

c. C

d. D

2.9i **Which of the following scientific names is a tautonym?**

a. *Canis lupus dingo*

b. *Suricata suricatta*

c. *Caracal caracal*

d. *Mus musculus*

2.10i **The family Ornithorhynchidae contains the**

a. echidnas

b. opossums

c. bandicoots

d. duck-billed platypus

2.11i **The order Rodentia contains the largest number of extant mammal species. Which mammalian order contains the second largest number of extant species?**

a. Primates

b. Chiroptera

c. Carnivora

d. Artiodactyla

2.12i **The subspecific names *troglodytes*, *schweinfurthii*, *vellerosus* and *verus* all refer to subspecies of**

a. gorillas

b. orangutans

c. bonobos

d. chimpanzees

2.13i **The hyoid bone is located in the**

    a. neck

    b. foot

    c. spine

    d. skull

2.14i **Rabbits and hares are currently classified in the order Lagomorpha, but until the early 20th century they were placed in the order**

    a. Perissodactyla

    b. Hyracoidea

    c. Artiodactyla

    d. Rodentia

2.15i **The puma (*Puma concolor*) has a large number of vernacular names because**

    a. it has a large number of subspecies

    b. it is the national animal of many countries within its range

    c. it has an extensive range inhabited by people who speak a large number of languages

    d. it has been reclassified several times

2.16i **Which of the following orders of mammals is no longer used in modern classifications?**

    a. Xenarthra

    b. Edentata

    c. Microbiotheria

    d. Eulipotyphla

2.17i **Which of the following statements about the blue whale (*Balaenoptera musculus*) is false?**

    a. It is the largest species of mammal to have ever inhabited the Earth

    b. It is a toothed whale

    c. It is a rorqual

    d. It occasionally hybridises with fin whales (*Balaenoptera physalus*)

2.18i   Figure 2.6 shows a model of a

    a. *Brontotherium*

    b. *Coelodonta*

    c. *Rhinoceros*

    d. *Elasmotherium*

Fig. 2.6.

2.19i   The Gomphotheriidae, Mammutidae and Moeritheriidae are extinct families of

    a. proboscideans

    b. artiodactyls

    c. perissodactyls

    c. sirenians

2.20i **Modern humans are most closely related to**

    i. bonobos

    ii. gorillas

    iii. orangutans

    iv. chimpanzees

    a. i and ii

    b. ii and iv

    c. i and iv

    d. iii and iv

Advanced

2.1a **Fossils of the diminutive hominins nicknamed 'hobbits' (*Homo floresiensis*) were found in 2004 in**

    a. Malta

    b. Sri Lanka

    c. Indonesia

    d. Japan

2.2a **The fossils known as the 'Laetoli footprints' were discovered in Tanzania in 1978. They are believed to have been made by**

    a. *Homo habilis*

    b. *Australopithecus africanus*

    c. *Homo erectus*

    d. *Australopithecus afarensis*

2.3a **Mark Leakey discovered the fossil hominin known as**

    a. *Zinjanthropus* (*Paranthropus*)

    b. *Proconsul*

    c. Lucy

    d. Ardi

2.4a   **'La Grande Coupure' refers to**

a. a period of massive expansion in the diversity of mammals

b. the time when mammals secondarily returned to the sea

c. a period when the ice caps expanded pushing mammalian species away from the poles

d. a wave of extinction of mammals that occurred about 32 million years ago

2.5a   **The tendency for mammalian lineages to increase in body size over geological time is known as**

a. Coope's rule

b. Cope's rule

c. Cooke's rule

d. Cole's rule

2.6a   **Which of the following is not a genus of fossil primates?**

a. *Mystacodon*

b. *Microcolobus*

c. *Afropithecus*

d. *Proconsul*

2.7a   **Within the Carnivora, the Caniformia is a**

a. family

b. subfamily

c. order

d. suborder

2.8a   **The monito del monte (*Dromiciops gliroides*) is the only species in the order**

a. Didelphimorphia

b   Microbiotheria

c. Notoryctemorphia

d. Soricomorpha

2.9a Where have fossils of woolly mammoths (*Mammuthus primigenius*) been found (Fig. 2.7)?

Fig. 2.7.

a. Africa and the Middle East

b. Southeast Asia

c. North America and Siberia

d. South and Central America

2.10a It is possible for palaeontologists who study mammals to determine their gestation period, how long they suckled for, and when they reached maturity by examining the chemistry of, and growth lines in, their

a. femurs

b. teeth

c. vertebrae

d. skulls

2.11a **The clade known as the Tethytheria contains the**

    a. dolphins, elephants and manatees

    b. elephants, dugongs and manatees

    c. dolphins, manatees and dugongs

    d. dugongs, porpoises and elephants

2.12a **The most distinctive characteristic of the primate group known as the Strepsirrhini is the possession of**

    a. flattened finger and toe nails

    b. forward-facing eyes

    c. a shortened snout

    d. a tooth comb made of lower incisors

2.13a **Farnsworth *et al.* (2023) have predicted that in 250 million years the continents will merge into one supercontinent and mammals will experience**

    a. a mass extinction caused by an increase in global temperatures

    b. a collapse in small mammal populations

    c. a collapse in marine mammal populations

    d. an increase in species diversity

2.14a **The closest extant relatives of the cetaceans are the**

    a. tapirs

    b. elephants

    c. hippopotamuses

    d. rhinoceroses

2.15a **Figure 2.8 shows a model of a member of the genus**

    a. *Entelodon*

    b. *Macrauchenia*

    c. *Uintatherium*

    d. *Hyaenodon*

Fig. 2.8.

2.16   The order Eulipotyphla includes the

a. moles and naked mole-rats

b. shrews and tree shrews

c. shrew opposums and colugos

d. hedgehogs, desmans and shrews

2.17a Which of the following families of bats is most speciose?

a. Phyllostomidae

b. Vespertilionidae

c. Rhinolophidae

d. Nycteridae

2.18a The mammalian clade Paenungulata consists of the orders

a. Proboscidea, Hyracoidea and Sirenia

b. Perrissodactyla and Artiodactyla

c. Sirenia and Cetacea

d. Lagomorpha, Rodentia and Hyracoidea

2.19a Which of the following is not a member of the Afrotheria?

   a. Proboscidea

   b. Tubulidentata

   c. Perissodactyla

   d. Sirenia

2.20a Which of the species shown in Fig. 2.9 is (are) not members of the Diprotodontia?

   a. i and iii

   b. ii and iv

   c. ii

   d. iv

Fig. 2.9.

3 Anatomy

This chapter contains questions concerned with the gross anatomy of mammals, especially their dentition and skeletal features.

Foundation

3.1f   What is the name of the structure labelled 'X' in Fig. 3.1?

Fig. 3.1.

    a. Coronoid process

    b. Mandibular foramen

    c. Diastema

    d. Foramen

© Paul A. Rees 2024. *Key Questions in Mammalogy* (P.A. Rees)
DOI: 10.1079/9781800624535.0003

3.2f How many bones are present in each side of the lower jaw of a mammal?

    a. 1

    b. 2

    c. 3

    d. 4

3.3f How many cervical vertebrae are possessed by most mammals?

    a. 5

    b. 6

    c. 7

    d. 8

3.4f 'Lordosis' is the name given to the natural curve of the

    a. top of the skull

    b. lumbar area of the spine

    c. lower jaw

    d. pelvis

3.5f Which of the following does not possess an omasum?

    a. Nubian ibex (*Capra nubiana*)

    b. Giraffe (*Giraffa camelopardalis*)

    c. Cape buffalo (*Syncerus caffer*)

    d. Asian elephant (*Elephas maximus*)

3.6f The baleen used to filter food from water in some whales is made of

    a. myosin

    b. bone

    c. keratin

    d. collagen

3.7f Which of the following are covered with protective scales made of keratin?

a. Pangolins

b. Echidnas

c. Tenrecs

d. Sengis

3.8f Mammals such as the skunk (*Mephitis mephitis*) and honey badger (*Mellivora capensis*) have distinctive black-and-white pelage (Fig. 3.2). The former produces foul-smelling liquid and the latter is extremely aggressive. It is possible that their distinctive markings warn off the predators of these species. This phenomenon is known as

a. astigmatism

b. aposematism

c. antagonism

d. agathism

Fig. 3.2.

3.9f   Which of the following felids cannot fully retract its claws?

    a. Tiger (*Panthera tigris*)

    b. Leopard (*Panthera pardus*)

    c. Cheetah (*Acinonyx jubatus*)

    d. Snow leopard (*Panthera uncia*)

3.10f Carnassial teeth are characteristic of

    a. carnivorans

    b. artiodactyls

    c. perissodactyls

    d. cetaceans

3.11f Fig. 3.3 is a stylised diagram of a mammalian heart. The structure labelled 'X' is the

    a. pulmonary vein

    b. pulmonary artery

    c. aorta

    d. superior vena cava

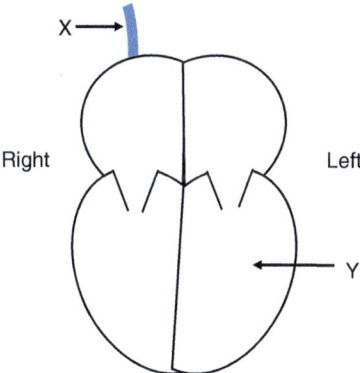

Fig. 3.3.

3.12f In Fig. 3.3 the cavity labelled 'Y' is the

    a. left atrium

    b. left septum

c. left ventricle

d. left auricle

3.13f Which of the following species has a vestigial tail?

a. Capybara (*Hydrochoerus hydrochaeris*)

b. Degu (*Octodon degus*)

c. Forest dormouse (*Dryomys nitedula*)

d. Nutria (*Myocastor coypus*)

3.14f Which of the following primates does not possess a prehensile tail?

a. Eastern black-and-white colobus (*Colobus guereza*)

b. Brown howler (*Alouatta guariba*)

c. Tufted capuchin (*Sapajus apella*)

d. Colombian spider monkey (*Ateles fusciceps*)

3.15f Purkinji fibres are found in the mammalian

a. brain

b. heart

c. skeletal muscles

d. smooth muscles

3.16f Some mammals possess nictitating membranes to protect their

a. ears

b. nose

c. mouth

d. eyes

3.17f Match the vernacular names of the animals in Table 3.1 with the images of their feet (Fig. 3.4).

a. A

b. B

c. C

d. D

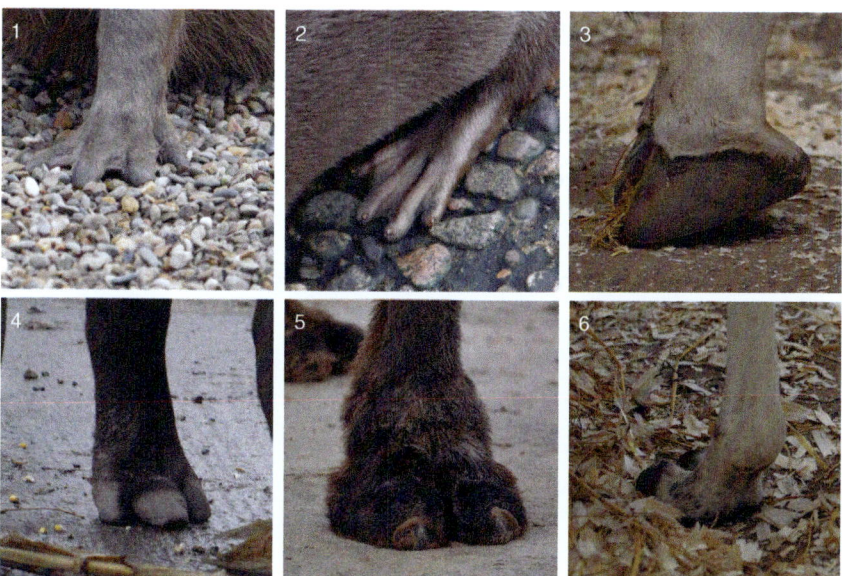

Fig. 3.4.

Table 3.1

Image	A	B	C	D
1	Capybara	Capybara	Asian short-clawed otter	Capybara
2	Asian short-clawed otter	Asian short-clawed otter	Capybara	Asian short-clawed otter
3	Bactrian camel	Bactrian camel	Llama	Giraffe
4	Brazilian tapir	Giraffe	Brazilian tapir	Brazilian tapir
5	Giraffe	Brazilian tapir	Bactrian camel	Bactrian camel
6	Llama	Llama	Giraffe	Llama

3.18f Which of the following possesses a patagium?

a. Matschie's tree-kangaroo (*Dendrolagus matschiei*)

b. Sumatran flying squirrel (*Hylopetes winstoni*)

c. Tasmanian devil (*Sarcophilus harrisii*)

d. Atlantic bottlenose dolphin (*Tursiops truncatus*)

3.19f **The turbinal bones are found in the**

    a. feet

    b. inner ear

    c. nasal chamber

    d. pelvic girdle

3.20f **The masseter muscle assists in moving the**

    a. jaw

    b. foot

    c. hand

    d. leg

Intermediate

3.1i **The foot of the species shown in Fig. 3.5 may be described as**

    a. bipedal

    b. plantigrade

    c. digitigrade

    d. unguligrade

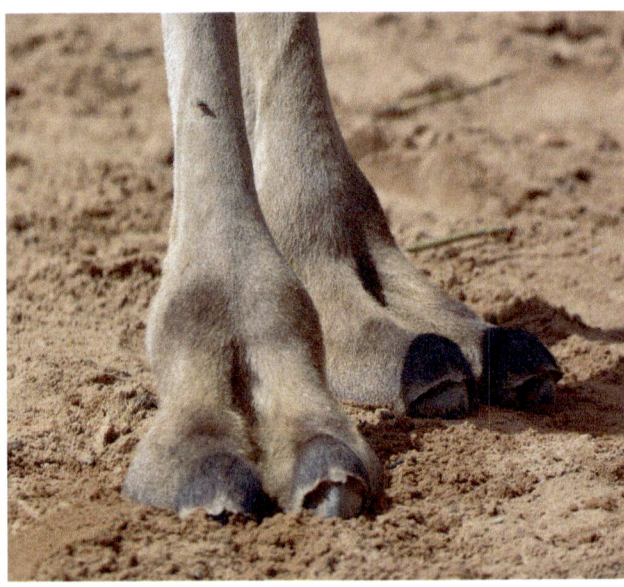

Fig. 3.5.

3.2i The limb shown in Fig. 3.6 belongs to a species that is a member of the

    a. Dermoptera

    b. Dasyuromorphia

    c. Scandentia

    d. Pilosa

Fig. 3.6.

3.3i Mammalian muscle is made up of the proteins

    a. actin and myosin

    b. myosin and fibrin

    c. actin and collagen

    d. fibrin and collagen

3.4i **Which of the animals shown in Fig. 3.7 possess ossicones?**

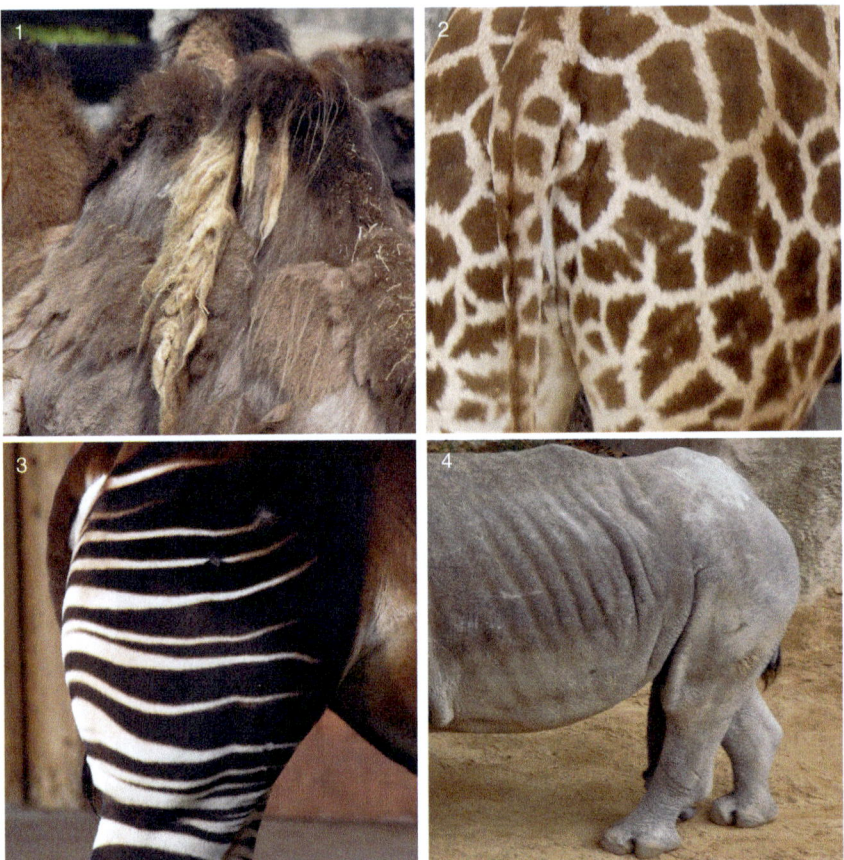

Fig. 3.7.

    a. 2 only

    b. 2 and 3

    c. 1 and 4

    d. 1, 2 and 3

3.5i **To which part of the skeleton of a whale does the structure shown in Fig. 3.8 belong?**

    a. Skull

    b. Pelvic girdle

    c. Vertebral column

    d. Mandible

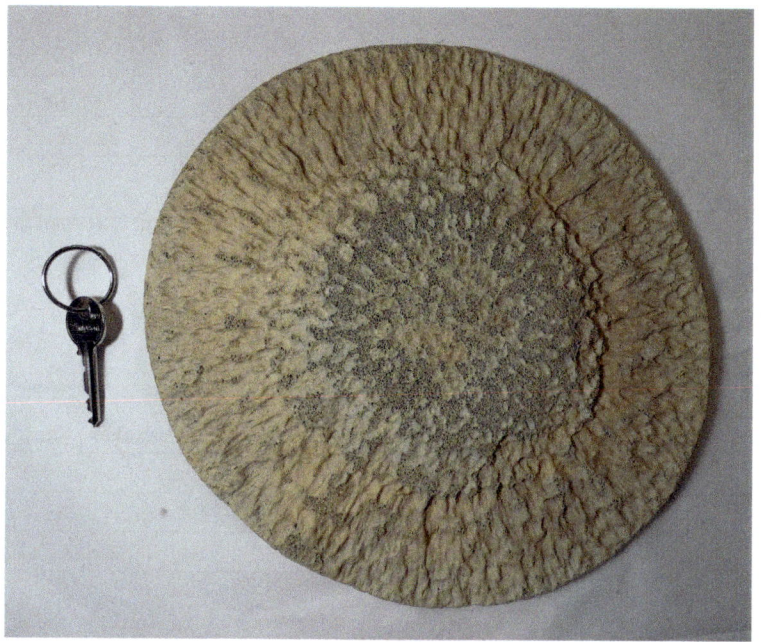

Fig. 3.8.

3.6i   The hand or forefoot of a mammal is called the

    a. pes

    b. hallux

    c. manus

    d. pollex

3.7i   The corpus callosum

    a. is part of the kidney

    b. connects the two sides of the brain

    c. is a fenestration in the skull

    d. is a feature of herbivore dentition

3.8i   In a typical terrestrial mammal, in which direction do the elbows and knees face (Table 3.2)?

    a. A

    b. B

    c. C

    d. D

Table 3.2

	A	B	C	D
Elbows	Backwards	Forwards	Forwards	Backwards
Knees	Forwards	Backwards	Forwards	Backwards

3.9i **Which of the following joints is responsible for the two head movements nodding and shaking (Table 3.3)?**

Table 3.3

	Head movement	
	Nodding	Shaking
A	Back of skull and atlas	Atlas and axis
B	Back of skull and axis	Atlas and axis
C	Atlas and axis	Back of skull and atlas
D	Atlas and axis	Back of skull and axis

    a. A

    b. B

    c. C

    d. D

3.10i **Which of the dental formulae below represents the dentition of the pika (*Ochotona dauurica*)? [Note: I = incisors; C = canines, P = premolars; M = molars.]**

    a. I 3/3   C 1/1   P 3/2   M 1/1

    b. I 3/3   C 1/1   P 4/4   M 2/3

    c. I 2/2   C 1/1   P 2/2   M 3/3

    d. I 2/1   C 0/0   P 3/2   M 2/3

3.11i **The structure labelled 'X' in Fig. 3.9 is the**

    a. occipital condyle

    b. foramen magnum

    c. lacuna

    d. foramen ovale

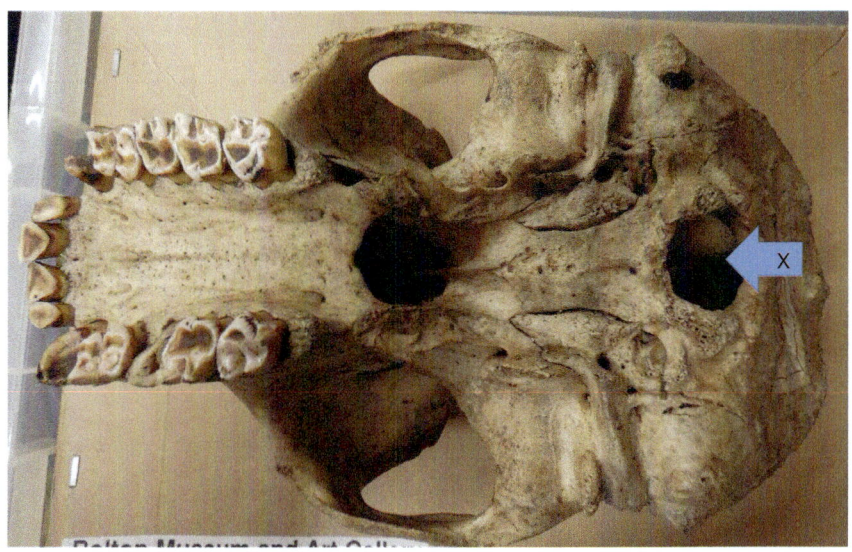

Fig. 3.9.

3.12i The tooth labelled 'X' in Fig. 3.10 is a

    a. protoloph tooth

    b. selenodont tooth

    c. metaloph tooth

    d. carnassial tooth

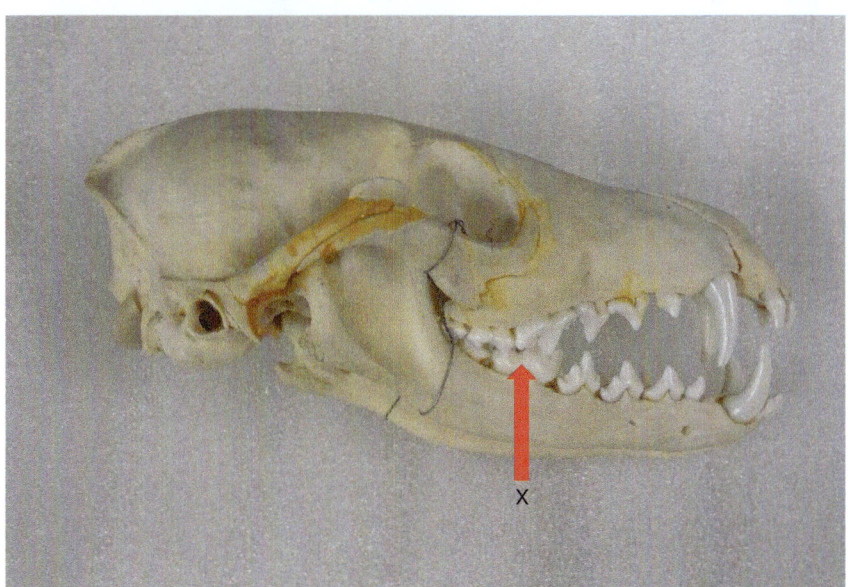

Fig. 3.10.

3.13i **Vibrissae are specialised hairs that**

    a. form spines in porcupines

    b. improve the insulation of the body

    c. contain special pigmentation for attracting mates

    d. have a sensory function

3.14i **The skull of a gorilla is shown in Fig. 3.11. Which of the structures labelled is the sagittal crest?**

    a. A

    b. B

    c. C

    d. D

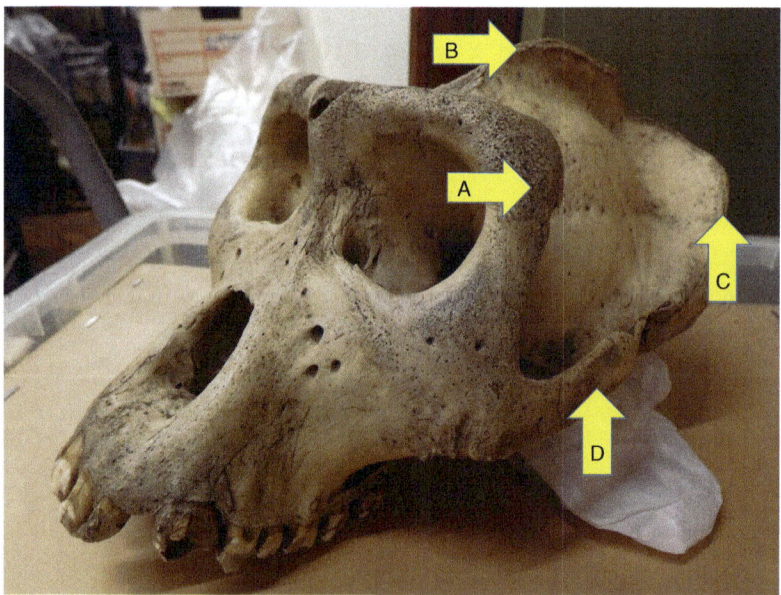

Fig. 3.11.

3.15i The structure in Fig. 3.12 is

    a. the baleen of a minke whale (*Balaenoptera*)

    b. part of the lining of the reticulum of a cow (*Bos*)

    c. the tongue of a manatee (*Trichechus*)

    d. the spleen of a hippopotamus (*Hippopotamus*)

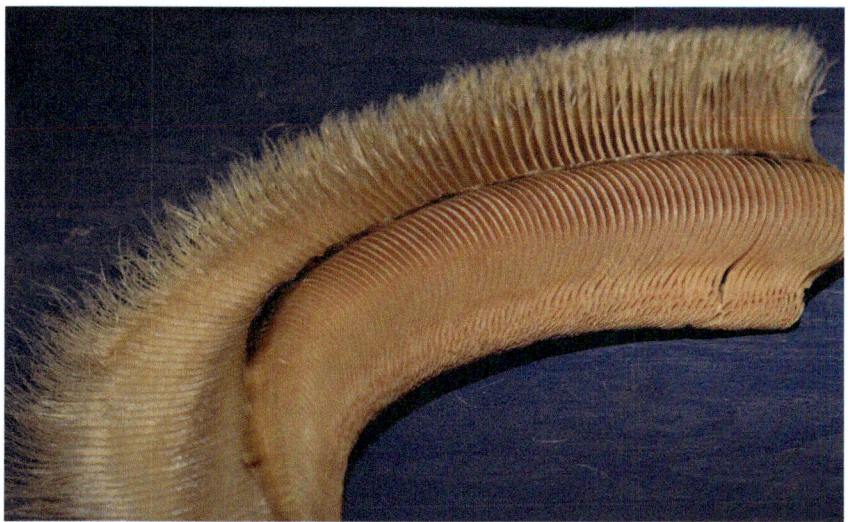

Fig. 3.12.

3.16i Females of which of the species shown in Fig. 3.13 possess(es) a pseudopenis?

    a. A   ii and iii

    b. B   i and ii

    c. C   i only

    d. D   ii and iv only

Fig. 3.13.

3.17i **The paired articulating surfaces at the base of the skull are the**

    a. occipital condyles

    b. ischial callosities

    c. alveolar processes

    d. styloid processes

3.18i **The primary purpose of the sagittal crest of the gorilla (*Gorilla spp.*) is to provide a surface for the attachment of the**

    a. mentalis muscle

    b. frontalis muscle

    c. triangularis muscle

    d. temporalis muscle

3.19i **The mammalian vertebral column consists of five sections which occur in which of the following sequences (Table 3.4)?**

Table 3.4

		A	B	C	D
Head end	▲	Cervical	Cervical	Caudal	Cervical
		Lumbar	Thoracic	Thoracic	Thoracic
		Thoracic	Lumbar	Lumbar	Lumbar
		Sacral	Sacral	Sacral	Caudal
Tail end	▼	Caudal	Caudal	Cervical	Sacral

a. A

b. B

c. C

d. D

3.20i **Ischial callosities are characteristically found in**

a. dolphins

b. elephants

c. baboons

d. giraffes

Advanced

3.1a **Match each skull in Fig. 3.14 with the correct species listed in Table 3.5**

a. A

b. B

c. C

d. D

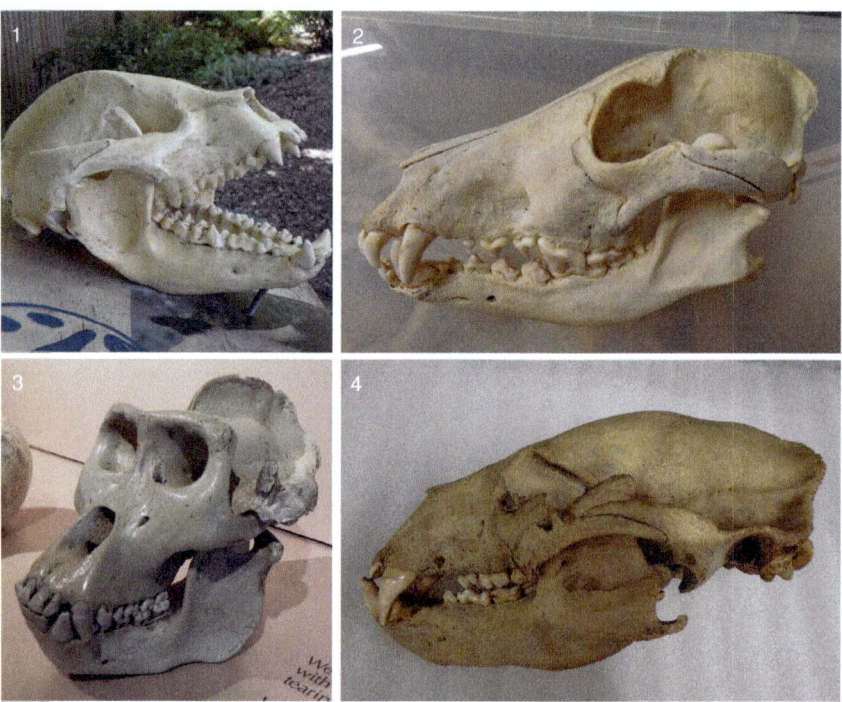

Fig. 3.14.

Table 3.5

Skull	A	B	C	D
1	Gorilla	Giant panda	Gorilla	Giant panda
2	Grey wolf	Grey wolf	Lion	Grey wolf
3	Giant panda	Gorilla	Grey wolf	Lion
4	Lion	Lion	Giant panda	Gorilla

3.2a **How many incisors are normally present in the upper jaw of an adult giraffe (*Giraffa camelopardalis*)?**

a. None

b. 2

c. 4

d. 6

3.3a **The dactylopatagium is part of the**

    a. foot of a rhinoceros

    b. wing of a bat

    c. tail of a porpoise

    d. ear of an elephant

3.4a **Cetaceans possess a complex network of blood vessels in their extremities that helps them to conserve heat using a counter-current exchange mechanism. This network is called a**

    a. vasa vasorum

    b. tunica externa

    c. vasa previa

    d. rete mirabile

3.5a **Which of the following do not normally exhibit syndactyly?**

    a. Kangaroos

    b. Siamangs

    c. Bears

    d. Wombats

3.6a **In mammals, what may be duplex, bipartite, bicornate or simplex?**

    a. The stomach

    b. The uterus

    c. The penis

    d. The caecum

3.7a **Which of the following species does not possess a baculum?**

    a. Human (*Homo sapiens*)

    b. Chimpanzee (*Pan troglodytes*)

    c. Walrus (*Odobenus rosmarus*)

    d. Fossa (*Cryptoprocta ferox*)

3.8a   Primates possess dermatoglyphics

    a. on their fingers

    b. in their stomachs

    c. on their retinas

    d. in their ears

3.9a   Figure 3.15 is a photograph of the skull of a lion (*Panthera leo*). Which of the labelled structures is the zygomatic arch?

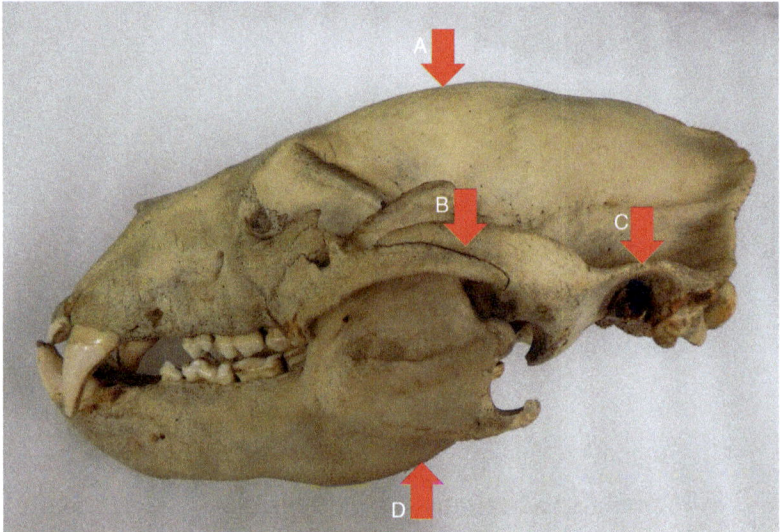

Fig. 3.15.

    a. A

    b. B

    c. C

    d. D

3.10a The yapock (yapok) (*Chironectes minimus*) is unique because

    a. it is a marsupial but does not have a pouch

    b. it always gives birth to twins

    c. it is capable of gliding

    d. it is the only extant marsupial in which both sexes have a pouch

3.11a **Whales find each other in the oceans by communicating using sound. According to Elemans *et al.* (2024) this sound is produced by**

    a. the larynx in all species

    b. a nasal organ in all species

    c. a nasal organ in baleen whales (mysticetes) and the larynx in toothed whales (odontocetes)

    d. the larynx in baleen whales (mysticetes) and a nasal organ in toothed whales (odontocetes)

3.12a **The fundus, cardia and pylorus are regions of the**

    a. stomach

    b. liver

    c. kidney

    d. uterus

3.13a **Who conducted a study of mammoth jaw bones and demonstrated that, although they were similar to those of African and Asian elephants, they were from a different species?**

    a. Jean-Baptiste Lamarck

    b. Charles Darwin

    c. Georges Cuvier

    d. Ernst Haeckel

3.14a **Which of the following species of lemur possesses unusually long fingers?**

    a. Alaotran gentle lemur (*Hapalemur alaotrensis*)

    b. Aye-aye (*Daubentonia madagascariensis*)

    c. Red-tailed sportive lemur (*Lepilemur ruficaudatus*)

    d. Diademed sifaka (*Propithecus diadema*)

3.15a **Individuals of which of the following species do not possess a cloaca?**

    a. Duck-billed platypus (*Ornithorhyncus anatinus*)

    b. Lesser hedgehog tenrec (*Echinops telfairi*)

    c. Short-beaked echidna (*Tachyglossus aculeatus*)

    d. All of the above possess a cloaca

3.16a **The type of dentition shown in Fig. 3.16 is known as**

    a. bunodont

    b. lophodont

    c. hypsodont

    d. brachydont

Fig. 3.16.

3.17a **The prezygapophysis and postzygapophysis are structures that form part of**

    a. some vertebrae

    b. the skull

    c. the femur

    d. the ilium

3.18a The Islets of Langerhans occur in the

    a. liver

    b. heart

    c. pancreas

    d. kidneys

3.19a Ruffini and Pacinian corpuscles are located in the

    a. ear

    b. mouth

    c. eye

    d. skin

3.20a The structure labelled 'X' in Fig. 3.17 is called a

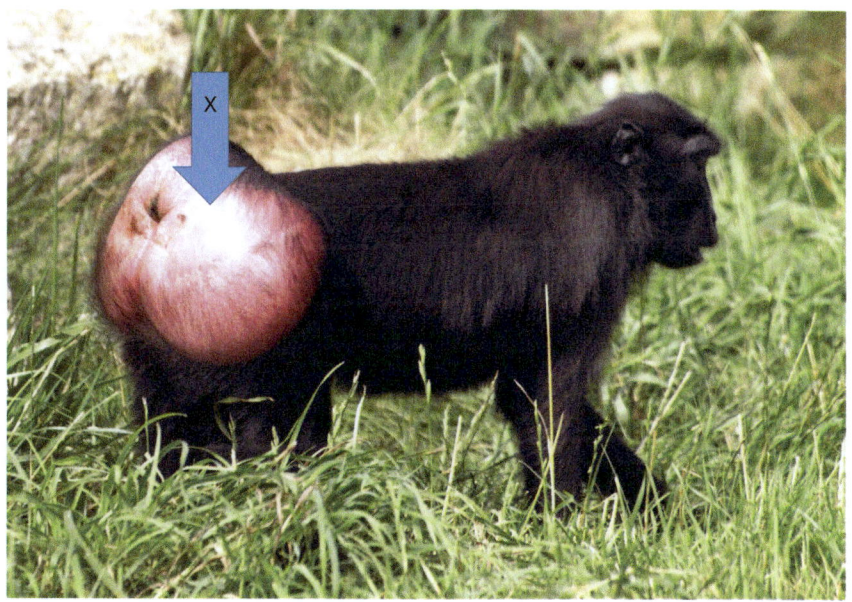

Fig. 3.17.

    a. perianal adornment

    b. perineal tumescence

    c. ischial tuberosity

    d. pelvic pubescence

4 Physiology

This chapter contains questions about the basic physiology of mammals, some of their unusual physiological characteristics and their adaptations to extreme environments.

Foundation

4.1f **Which of the following animals employ saltatory (saltatorial) locomotion?**

   a. Elephant

   b. Dolphin

   c. Wallaby

   d. Antelope

4.2f **In the higher primates the most important means of sensing the environment is the sense of**

   a. smell

   b. sight

   c. hearing

   d. touch

© Paul A. Rees 2024. *Key Questions in Mammalogy* (P.A. Rees)
DOI: 10.1079/9781800624535.0004

4.3f Which of the following mammals is most likely to have a resting heart rate of 1,200 beats per minute?

a. A tapir

b. A dolphin

c. An elephant

d. A shrew

4.4f Mature mammalian red blood cells

a. appear as biconcave discs

b. possess no nucleus

c. contain haemoglobin

d. have all of the above characteristics

4.5f The Malpighian body occurs in the kidney and is

a. another name for the glomerulus

b. another name for the Bowman's capsule

c. made up of the glomerulus and the Bowman's capsule

d. made up of the Bowman's capsule and the loop of Henlé

4.6f The structure that prevents food from entering the trachea during swallowing is called the

a. epiglottis

b. epistaxis

c. epineurium

d. epiphysis

4.7f Gestation is the period of development of a mammalian embryo between

a. mating and birth

b. conception and the beginning of labour

c. conception and birth

d. mating and the beginning of labour

4.8f **Match the species listed in Table 4.1 with the correct gestation period**

Table 4.1

Species	Gestation period (days)			
	A	B	C	D
Chimpanzee (*Pan troglodytes*)	280	236	108	236
Lion (*Panthera leo*)	108	108	63	280
Bison (*Bison bison*)	236	280	280	108
Coyote (*Canis latrans*)	63	63	236	63

a. A

b. B

c. C

d. D

4.9f **Physiological changes occur in many terrestrial mammals when they are submerged in water. These changes are collectively called the**

a. submergence reflex

b. diving reaction

c. diving reflex

d. diving adaptation

4.10f **In mammals the temperature of the blood is monitored by specialised cells in the**

a. hypothalamus

b. pons

c. cerebrum

d. thalamus

4.11f **Which muscles cause the hairs that make up mammalian fur to stand up when the body is exposed to cold?**

a. Adductor longus

b. Extensor retinaculum

    c. Rectus femoris

    d. Arrector pili

4.12f **Rock hyraxes (*Procavia capensis*) bask with their bodies broadside to the sun when it is low in the sky and huddle together on cold days (Fig. 4.1). This method of regulating body temperature is known as**

Fig. 4.1.

    a. behavioural thermoregulation

    b. postural thermogenesis

    c. orientation endothermy

    d. behavioural homiothermy

4.13f **Which of the following produces cuboidal faeces?**

    a. Ursine tree-kangaroo (*Dendrolagus ursinus*)

    b. Long-eared jerboa (*Euchoreutes naso*)

    c. Bare-nosed wombat (*Vombatus ursinus*)

    d. Eastern long-beaked echidna (*Zaglossus bartoni*)

4.14f **The epididymis**

    a. stores sperm cells

    b. filters toxins from the blood

    c. produces skin cells

    d. secretes oestrogen

4.15f **The composition of the milk of reindeer and marine mammals differs from that of humans and cows in containing more**

    a. carbohydrate

    b. fat

    c. calcium

    d. protein

4.16f **Which of the following types of glands produce a lubricant that protects and waterproofs fur?**

    a. sebaceous glands

    b. sweat glands

    c. endocrine glands

    d. parotid glands

4.17f **The rate of energy expenditure at rest is known as the**

    a. basal metabolic rate

    b. basal physiological rate

    c. minimum metabolic rate

    d. essential metabolic rate

4.18f **Which of the following is not true of koalas (*Phascolarctos cinereus*) (Fig. 4.2)?**

Fig. 4.2.

    a. They are able to detoxify some substances that are toxic to other mammals

    b. They have a similar metabolic rate to other mammals of the same size

    c. They spend about 20 hours asleep each day

    d. They possess a large caecum

4.19f **The normal blood pressure (systole/diastole) at the heart in a giraffe (*Giraffa camelopardalis*) is approximately**

    a. 110/70 mmHg

    b. 130/95 mmHg

    c. 220/180 mmHg

    d. 350/290 mmHg

4.20f The nasal passages of a dehydrated camel desaturate exhaled air. This is achieved by absorbing water using a layer of dried mucus and cellular debris that coats the nasal passage and is

    a. hydroscopic

    b. hydrophobic

    c. hydrophilic

    d. hygroscopic

Intermediate

4.1i Which of the following species has kidneys with an exceptionally long Loop of Henlé?

    a. Asian elephant (*Elephas maximus*)

    b. Fallow deer (*Dama dama*)

    c. Kangaroo rat (*Dipodomys*)

    d. Eastern grey kangaroo (*Macropus giganteus*)

4.2i Mammalian milk contains the sugar lactose which is made up of

    a. galactose and fructose

    b. glucose and fructose

    c. maltose and fructose

    d. glucose and galactose

4.3i In mammalian physiology the term 'diastole' refers to an action of the

    a. brain

    b. lungs

    c. heart

    d. kidneys

4.4i Cold-adapted mammals often offset the energetic costs of keeping warm by allowing the temperature of their skin or extremities to fall well below their core body temperature. This is practice is known as

a. adaptive homiothermy

b. adaptive heterothermy

c. adaptive thermogenesis

d. adaptive vasoconstriction

4.5i Prolactin is a mammalian hormone that controls

a. gastric secretion

b. milk production

c. ovulation

d. glucose metabolism

4.6i The Jacobson's organ found in some mammalian taxa is an organ of

a. vision

b. hearing

c. mechanoreception

d. chemoreception

4.7i Which of the following stores energy and increases the efficiency of running in ungulates?

a. A tendon and ligament spring

b. A muscle and tendon spring

c. A ligament and muscle spring

d. A muscle and cartilage spring

4.8i Williams *et al.* (2016) studied the ecophysiology of the Arctic ground squirrel (*Urocitellus parryii*) in free-living conditions using biologgers. Which of the following statements about the mechanisms used by this species to allow it to survive in Arctic conditions is false?

a. Allowing its core body temperature to drop below 10°C during torpor

b. Having torpor bouts that last several weeks

c. Reducing its metabolic rate to 1–2% of basal metabolism during torpor

d. Every 10–21 days it spontaneously rewarms to euthermic levels (36 – 37°C) for about a day before re-entering torpor

4.9i Ovulation in mammals follows a surge in the blood level of

a. cholecystokinin

b. thyroxine

c. parathyroid hormone

d. luteinising hormone

4.10i Meissner's corpuscles are specialised nerve endings that are only found in

a. cetaceans

b. primates

c. rodents

d. bats

4.11i In mammals, primer, releaser, signaller and modulator are all types of

a. enzymes

b. hormones

c. pheromones

d. neurotransmitters

4.12i **Which of the following has the lowest metabolic rate recorded in any mammal?**

    a. Asian elephant (*Elephas maximus*)

    b. Three-toed sloth (*Bradypus*)

    c. Aardvark (*Orycteropus afer*)

    d. Nine-banded armadillo (*Dasypus novemcinctus*)

4.13i **Which of the following is the only species of primate that is venomous?**

    a. Pygmy slow loris (*Xanthonycticebus pygmaeus*)

    b. Greater dwarf lemur (*Cheirogaleus major*)

    c. Central African potto (*Perodicticus edwardsi*)

    d. Garnett's galago (*Otolemur garnettii*)

4.14i **The corpus luteum is a gland involved in**

    a. digestion

    b. reproduction

    c. excretion

    d. lactation

4.15i **Young monotremes hatch at a relatively early stage of their development so are highly dependent on a parent for their survival. These animals are said to be**

    a. altricial

    b. precocial

    c. autonomous

    d. nidifugous

4.16i **Pheromones have been most extensively studied in**

    a. dogs

    b. rabbits

    c. rats

    d. house mice

4.17i Fig. 4.3 is a stylised diagram of part of a mammalian circulatory system. In this figure the blood vessel labelled 'X' is the

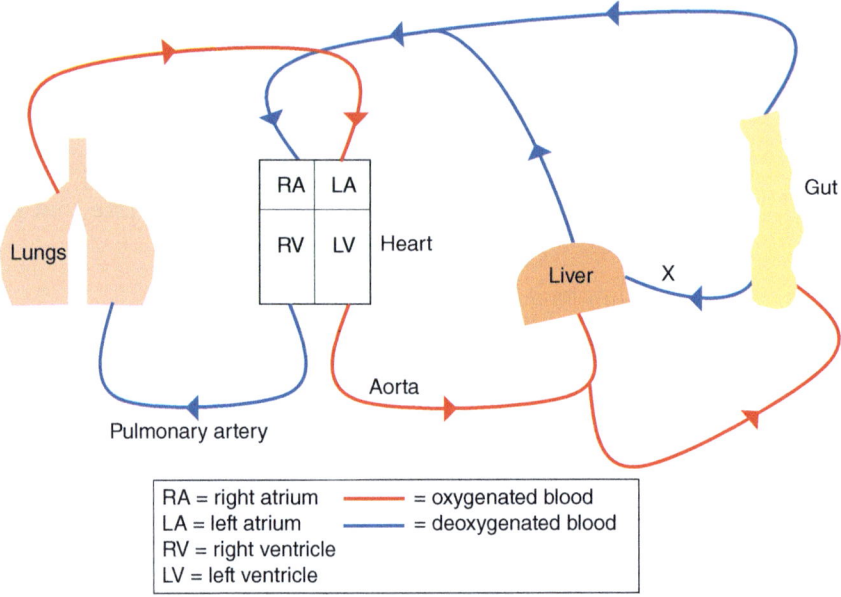

Fig. 4.3.

    a. hepatic artery

    b. hepatic vein

    c. hepatic portal vein

    d. mesenteric vein

4.18i Which of the following hormones induces uterine contractions during labour and the release of milk from mammary glands?

    a. Progesterone

    b. Epinephrine

    c. Aldosterone

    d. Oxytocin

4.19i The HPA axis plays an important role in

    a. the body's reaction to stress

    b. maintaining the implanted embryo

   c. fighting disease

   d. growth and development

4.20i **Golden moles (Chrysochloridae) live in arid regions where food and water are scarce. They have adapted to this during the course of evolution by**

   a. having a low metabolic rate

   b. having a low body temperature

   c. entering a state of torpor for parts of most days

   d. doing all of the above

Advanced

4.1a **In winter, the common shrew (*Sorex araneus*) saves energy by shrinking the size of its skull. In spring it regrows it again. This phenomenon was discovered in 1949 by**

   a. August Gustaw Dehnel

   b. Ernst Haeckel

   c. Robert Hinde

   d. Jean Dorst

4.2a **The network of blood vessels that assists in cooling the blood passing to the brain of a gazelle is called the**

   a. jugular rete

   b. carotid rete

   c. cranial rete

   d. cerebral rete

4.3a Which row in Table 4.2 most accurately resembles the relationship between the red blood cell numbers in the blood and the myoglobin concentration in the muscles of marine mammals compared with terrestrial mammals?

Table 4.2

	Red cell numbers in blood		Myoglobin concentration in muscles	
	Terrestrial mammal	Marine mammal	Terrestrial mammal	Marine mammal
A	1	1	1	9
B	1	1	1	20
C	1	2	1	9
D	1	2	1	20

    a. A

    b. B

    c. C

    d. D

4.4a Which of the following exhibit semelparity?

    a. Antechinuses (marsupial mice)

    b. Chevrotains (mouse-deer)

    c. Tamarins

    d. Tamanduas

4.5a Eimer's organ is a specialised touch receptor located on the

    a. feet of macropods

    b. beaks of dolphins and porpoises

    c. snouts of moles and desmans

    d. fingers of primates

4.6a The stridulating organ of the tenrec (*Hemicentetes*) consists of a group of specialised quills on the middle of the back which are used in

a. defence

b. communication

c. mating

d. parental care

4.7a The forelimb-hindlimb phase relationship is important in understanding

a. gait in quadrupeds

b. the development of the axial skeleton in the mammalian embryo

c. diving in cetaceans

d. brachiation in gibbons

4.8a Match the times taken to enter torpor listed in Table 4.3 with the correct taxon.

Table 4.3

Taxon	Time taken to enter torpor (mins)			
	A	B	C	D
Suncus (shrew)	8307	35	35	35
Tachyglossus (echidna)	2685	1648	2685	2685
Taxidea (badger)	1648	2685	8307	1648
Ursus (bear)	35	8307	1648	8307

a. A

b. B

c. C

d. D

4.9a In a sperm whale (*Physeter macrocephalus*) the spermaceti organ is used to

a. produce spermatozoa

b. detoxify the blood

c. control water loss

d. dissipate excess heat

4.10a **Figure 4.4 shows the relationship between the partial pressure of oxygen and the oxygen saturation of haemoglobin in the blood as an oxygen dissociation curve. As the concentration of carbon dioxide in the blood increases, for example in respiring tissues, the blood becomes more acidic. This increase in acidity reduces the affinity of the haemoglobin for oxygen so the blood gives up its oxygen more readily. The result is a shift of the oxygen association curve to the right. This is know as the**

a. Baumer effect

b. Baur effect

c. Bohr effect

d. Bauch effect

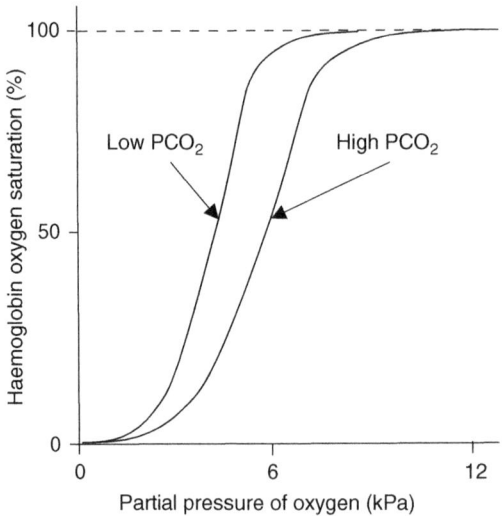

Fig. 4.4.

4.11a **Which of the options in Table 4.4 shows the correct sequence of the stages in the oestrous cycle?**

Table 4.4

Stage	A	B	C	D
First	Pro-oestrus	Metoestrus	Ovulation	Pro-oestrus
Second	Dioestrus	Ovulation	Pro-oestrus	Ovulation
Third	Ovulation	Pro-oestrus	Metoestrus	Metoestrus
Fourth	Metoestrus	Dioestrus	Dioestrus	Dioestrus

     a. A

     b. B

     c. C

     d. D

4.12a **A multiparous mammal has had many**

     a. diseases

     b. mates

     c. miscarriages

     d. offspring

4.13a **Macropods (Fig. 4.5) may experience a period of arrested development of the blastocyst known as**

Fig. 4.5.

a. embryonic suspension

b. embryonic dormancy

c. embryonic diapause

d. developmental diapause

4.14a Fig. 4.6 shows an oxygen dissociation curve for haemoglobin and four other lines (A, B, C and D). Which of these lines shows the oxygen dissociation curve of myoglobin?

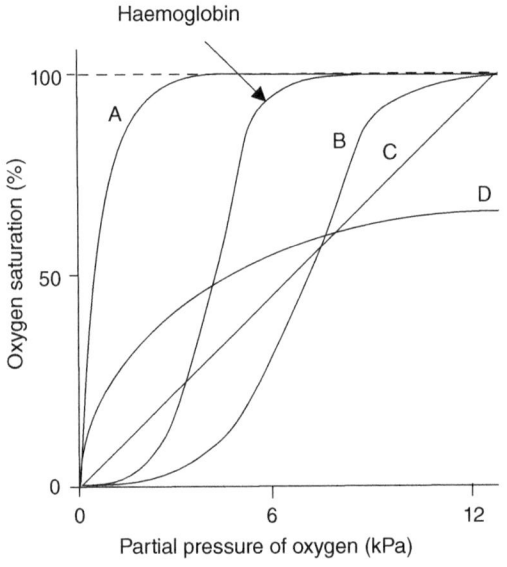

Fig. 4.6.

a. A

b. B

c. C

d. D

4.15a Deciduate mammals

a. shed the uterine lining with the foetal membranes as the 'afterbirth' at parturition

b. only produce a single offspring at a time

c. possess milk teeth that are shed and replaced by permanent teeth

d. are only able to digest the leaves and bark of deciduous trees

4.16a Which of the arrows in Fig. 4.7 indicates the source of the venom produced by the crural glands of a duck-billed platypus (*Ornithorhyncus anatinus*)?

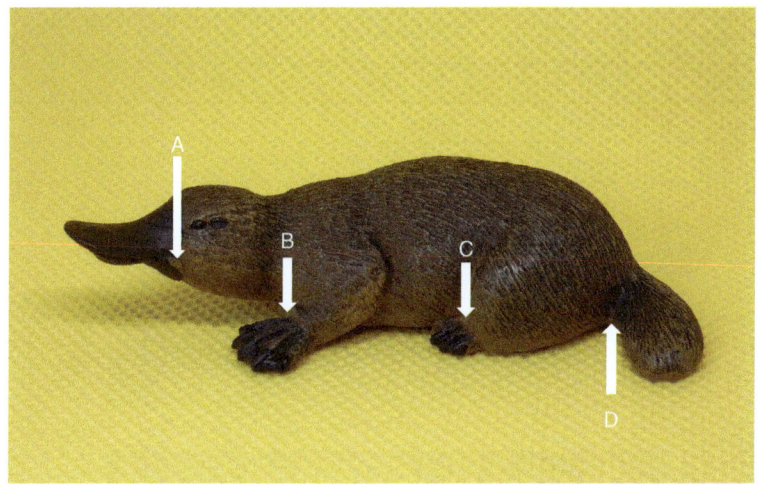

Fig. 4.7.

    a. A

    b. B

    c. C

    d. D

4.17a Members of which of the following taxa exhibit unihemispheric slow wave sleep (USWS) and a negligible amount or complete absence of rapid eye movement (REM) sleep?

    a. Proboscidea

    b. Artiodactyla

    c. Cetacea

    d. Sirenia

4.18a Which of the following mammals injects its prey with venom using its special grooved lower incisors?

    a. Solenodons

    b. Tenrecs

    c. Desmans

    d. Gymnures

4.19a The hippopotamus (*Hippopotamus amphibius*) (Fig. 4.8) 'sweats' two coloured pigments. Match the pigment colour to its function from the options in Table 4.5.

Table 4.5

Function	A	B	C	D
Antibiotic	Yellow	Red	Green	Orange
Sunscreen	Green	Orange	Yellow	Red

    a. A

    b. B

    c. C

    d. D

Fig. 4.8.

4.20a The *tapetum lucidum* is

    a. a structure in the ear that assists in the detection of high frequency sounds in bats

    b. a thin membrane that protects the eyes of marine mammals during diving

    c. a structure in the female ovary the produces hormones that regulate the oestrous cycle

    d. a structure in the eye that enhances visual sensitivity in low light levels

5 Behaviour

This chapter contains questions concerned with the behaviour of mammals in the wild and their behaviour in captivity. This includes feeding, reproductive and social behaviours, communication and cognition. It also contains questions on the biological basis of behaviour: neurobiology, sense organs and aspects of the sensory abilities of mammals.

Foundation

5.1f   Which of the following is not a type of affiliative behaviour in chimpanzees (*Pan troglodytes*)?

   a. Grooming

   b. Facial threat

   c. Food sharing

   d. Playing

5.2f   A large social group of spotted hyenas (*Crocuta crocuta*) is known as a

   a. gang

   b. tribe

   c. clan

   d. band

© Paul A. Rees 2024. *Key Questions in Mammalogy* (P.A. Rees)
DOI: 10.1079/9781800624535.0005

5.3f **The broad-headed spiny rat (*Clyomys laticeps*) is a semi-fossorial rodent. The term 'fossorial' refers to their habitat of**

 a. burrowing underground

 b. feeding in trees

 c. living among rocks

 d. hibernating in cold conditions

5.4f **The species shown in Fig. 5.1 lives in**

 a. pairs

 b. small groups of related individuals

 c. single-sex groups

 d. colonies

Fig. 5.1.

5.5f **In most species of social mammals the majority of dispersers are**

 a. juvenile males

 b. juvenile females

 c. adult males

 d. adult females

5.6f A field biologist plans to study chimpanzees (*Pan troglodytes*) in a forest in western Tanzania. She intends to construct an activity budget for each individual that will estimate the amount of time each spends on various activities such as walking, resting, sleeping, foraging, eating and so on. To do this she first needs to construct a document that carefully defines each of these activities. This is called an

    a. ecograph

    b. ecogram

    c. ethograph

    d. ethogram

5.7f Appeasement behaviour in social mammals often involves the appeaser

    a. lowering the body to the ground

    b. exposing vulnerable parts of the body, especially the abdomen

    c. males performing female behaviour, for example presenting the rump to the aggressor

    d. exhibiting some or all of the behaviours above

5.8f Mammals are capable of a much wider range of behaviours than other vertebrates and have proportionately larger brains. How much larger is the average mammalian brain compared with that of a reptile of similar size?

    a. x2

    b. x5

    c. x10

    d. x20

5.9f The Iberian wolves (*Canis lupus signatus*) in Fig. 5.2 live in a zoo and repeatedly pace along the same path in their enclosure for no apparent reason. This is known as a

    a. compulsive behaviour

    b. stereotypic behaviour

    c. appetitive behaviour

    d. fixed action pattern

Fig. 5.2.

5.10f **'Blood-sharing' is a behaviour observed in some members of the**

   a. Eulipotyphla

   b. Rodentia

   c. Chiroptera

   d. Cingulata

5.11f **Mammals possess sophisticated sensory systems. Match each structure in Table 5.1 with the sensory system of which it is a part.**

   a. A

   b. B

   c. C

   d. D

Table 5.1

Sensory system	A	B	C	D
Auditory	Fovea	Organ of Corti	Pacinian corpuscules	Jacobson's organ
Olfactory	Pacinian corpuscules	Jacobson's organ	Organ of Corti	Organ of Corti
Visual	Organ of Corti	Fovea	Jacobson's organ	Fovea
Tactile	Jacobson's organ	Pacinian corpuscules	Fovea	Pacinian corpuscules

5.12f Which aspect of primate behaviour is sometimes described as 'social cement' or 'social glue'?

a. Feeding

b. Grooming

c. Mating

d. Playing

5.13f Which of the following mammals may be described as eusocial?

a. Naked mole-rat (*Heterocephalus glaber*)

b. African elephant (*Loxodonta africana*)

c. Tiger (*Panthera tigris*)

d. Lesser cane rat (*Thryonomys gregorianus*)

5.14f Which of the species shown in Fig. 5.3 engages in an intense greeting ceremony involving nosing, lip licking, tail wagging and circling prior to hunting?

a. A

b. B

c. C

d. D

Fig. 5.3.

5.15f Groups of which of the following species adopt a defensive formation when they feel threatened, for example by predators?

a. Musk ox (*Ovibos moschatus*)

b. African elephant (*Loxodonta africana*)

c. North American bison (*Bison bison*)

d. All of the above

5.16f Species belonging to which of the following families have been observed using tools?

a. Elephantidae

b. Mustelidae

c. Delphinidae

d. All of the above

5.17f When a juvenile bull Asian elephant (*Elephas maximus*) observes adults mating the young bull may mount a juvenile cow. This may happen within seconds of observing the adults' behaviour. This phenomenon is called

a. imprinting

b. anticipatory behaviour

c. social facilitation

d. operant conditioning

5.18f Individuals of which of the following species engages in 'pronking' or 'stotting' when alarmed?

a. Springbok (*Antidocas marsupialis*)

b. Fallow deer (*Dama dama*)

c. Waterbuck (*Kobus ellipsiprymnus*)

d. Spanish ibex (*Capra pyrenaica*)

5.19f A rapid sequence of clicks produced by a whale or dolphin is called

a. a click sequence

b. a click train

c. a clack sequence

d. a clink array

5.20f The most heavily used area within a home range – often containing a nest, feeding site, water and other resources – is known as the

a. primary area

b. vital area

c. critical area

d. core area

Intermediate

5.1i **Which species did Harry Harlow use in his laboratory experiments on maternal separation and social deprivation in primates?**

    a. Rhesus monkeys (*Macaca mulatta*)

    b. Chimpanzees (*Pan troglodytes*)

    c. Squirrel monkeys (*Saimiri* spp.)

    d. Vervet monkeys (*Chlorocebus pygerythrus*)

5.2i **Female geladas (*Theropithecus gelada*) remain with their natal group. This behaviour is known as**

    a. paraphyly

    b. parapatry

    c. philopatry

    d. phylogeny

5.3i **Elephants extract salt to supplement their diets from caves in which of the following mountains?**

    a. Mt Elgon

    b. Mt Meru

    c. Mt Kenya

    d. Mt Kilimanjaro

5.4i **Which of the animals shown in Fig. 5.4 engages in 'stink fights'?**

    a. A

    b. B

    c. C

    d. D

Fig. 5.4.

5.5i The structure in a dolphin's head that focuses clicking sounds thereby acting as a type of acoustic lens that assists in sound recognition is called the

a. pineapple

b. melon

c. apple

d. peach

5.6i Bubble-net feeding is employed by

a. bottlenose dolphins (*Tursiops* spp.)

b. minke whales (*Balaenoptera acutorostrata*)

c. harbour seals (*Phoca vitulina*)

d. humpback whales (*Megaptera novaeangliae*)

5.7i **The activity of the Mayotte lemur (*Eulemur fulvus fulvus*) is distributed fairly evenly throughout a 24-hour cycle. This animal is considered to be**

a. vespertine

b. cathemeral

c. matutinal

d. crepuscular

5.8i **In mammals, chemical signals that elicit a specific and immediate behavioural effect are called**

a. releaser pheromones

b. reception pheromones

c. reaction pheromones

d. expression pheromones

5.9i **Young shrews are moved by their mother in a line whereby she is at the front and the young form a line in which each individual holds the base of the tail of the preceding shrew in its mouth. This line of shrews is known as a**

a. procession

b. conga

c. caravan

d. parade

5.10i **Match each sound in Table 5.2 with the type of animal that makes it.**

a. A

b. B

c. C

d. D

Table 5.2

Sound	Animal type			
	A	B	C	D
Feeding buzz	Bat	Chimpanzee	Bat	Dolphin
Low frequency rumble	Elephant	Dolphin	Dolphin	Elephant
Panted grunt	Chimpanzee	Bat	Elephant	Chimpanzee
Whistle	Dolphin	Elephant	Chimpanzee	Bat

5.11i An activity pattern based on a 24-hour cycle is known as a

    a. circannual rhythm

    b. circadian rhythm

    c. crepuscular rhythm

    d. diurnal rhythm

5.12i Which of the following species does not engage in cooperative hunting?

    a. Killer whale (*Orcinus orca*)

    b. Chimpanzee (*Pan troglodytes*)

    c. Bottlenose dolphin (*Tursiops truncatus*)

    d. Maned wolf (*Chrysocyon brachyurus*)

5.13i The linguistic and cognitive abilities of *Kanzi* and *Panbanisha* have been studied in detail by Dr Sue Savage-Rumbaugh. These animals were

    a. Sumatran orangutans (*Pongo abelii*)

    b. chimpanzees (*Pan troglodytes*)

    c. Western lowland gorillas (*Gorilla gorilla gorilla*)

    d. bonobos (*Pan paniscus*)

5.14i A scansorial mammal has the ability to

    a. run

    b. climb

c. swim

d. jump

5.15i **Which species of seal is able to inflate a black structure on the top of its head and the lining of its left nostril into a red bladder as part of its mating display?**

a. Bearded seal (*Erignathus barbatus*)

b. Baikal seal (*Pusa sibirica*)

c. Hooded seal (*Cystophora cristata*)

d. Leopard seal (*Hydrurga leptonyx*)

5.16i **Bottlenose dolphins (*Tursiops*) have a fission-fusion social system whereby**

a. juvenile males disperse from their natal group

b. juvenile females disperse from their natal group

c. individuals come together to form large, stable groups

d. members of small groups move freely between groups

5.17i **Which of the following are examples of alloparenting?**

i. A male grey wolf (*Canis lupus*) regurgitating food for his pups

ii. A female elephant allowing an unrelated calf to suckle

iii. A young Bornean orangutan (*Pongo pygmaeus*) learning which fruits to eat by watching his mother

iv. A human female caring for a child that is not her own while the mother goes to work

v. An adult female chimpanzee (*Pan troglodytes*) carrying her daughter's infant

a. i only

b. i and iii

c. ii, iv and v

d. i, ii and iii

5.18i **In all species of mammals at least one parent provides at least some care for their offspring (Fig. 5.5). A behaviour pattern that increases the offspring's chances of survival at the expense of the parents' ability to rear future offspring is known as**

Fig. 5.5.

    a. parental sacrifice

    b. patriarchal expenditure

    c. matriarchal commitment

    d. parental investment

5.19i **Payne *et al.* (1986) were the first to report the use of infrasound (frequencies below the threshold of human hearing) in communication between individuals of which of the following?**

    a. African elephants (*Loxodonta africana*) in Amboseli National Park, Kenya

    b. Asian elephants (*Elephas maximus*) at the Washington Park Zoo, Portland, Oregon

    c. African elephants (*Loxodonta africana*) in Hwange National Park, Zimbabwe

    d. Asian elephants (*Elephas maximus*) in Yala National Park, Sri Lanka

5.20i Which of the following types of spacing calls are used by primates (Fig. 5.6)?

Fig. 5.6.

    i.   Distance-increasing signals used to drive other groups away

    ii. Distance-maintaining signals which regulate the use of overlapping ranges

    iii. Distance-reducing signals such as 'contact calls' and 'lost calls'

    iv. Proximity-maintaining signals used by individuals during social grooming within groups

    a. i, ii and iv

    b. ii, iii and iv

    c. iii and iv

    d. i, ii, iii and iv

Advanced

5.1a   In some species of voles, pregnant females abort their foe-
tuses in response to the presence of a new male (or his
odour), enter oestrus and then breed with the new male
within days. This male-induced termination is known as the

    a. Butler effect

    b. Brant effect

    c. Bronson effect

    d. Bruce effect

5.2a   A laufschlag (or leg beat) is a behaviour performed by roan
antelopes (*Hippotragus equinus*) (Fig. 5.7) as a behavioural
element of

Fig. 5.7.

    a. aggression

    b. appeasement

    c. courtship

    d. feeding

5.3a An adult male of which of the following species was observed in Indonesia self-medicating by making a paste from the chewed leaves and stem of a plant known to have anti-bacterial and anti-inflammatory properties, and then smearing it on a wound on his face for a period of several days after which the wound had healed?

  a. A Sumatran orangutan (*Pongo abelii*)

  b. A Bornean gibbon (*Hylobates muelleri*)

  c. A West Javan langur (*Trachypithecus mauritius*)

  d. An Asian elephant (*Elephas maximus*)

5.4a Packer and Pusey (1982) studied coalitions of male lions in the Serengeti and used DNA fingerprinting to establish that

  a. coalitions only contained related males

  b. coalitions only contained unrelated males

  c. approximately 10 per cent of coalitions included at least one unrelated male

  d. approximately 50 per cent of coalitions included at least one unrelated male

5.5a A study of cooperation in rats employed a mathematical game known as the

  a. Prisoner's Dilemma

  b. Hostage's Quandary

  c. Convict's Predicament

  d. Captive's Difficulty

5.6a In red deer (*Cervus elaphus*) rutting behaviour is triggered by a change in

  a. light intensity

  b. day length

  c. food quality

  d. temperature

5.7a *Happy*, a resident at the Bronx Zoo in New York, demonstrated that she was self-aware by touching an 'X' painted above her eye when she could see it in a large mirror. *Happy* was

a. a gorilla

b. an elephant

c. a chimpanzee

d. a bonobo

5.8a A review of studies of changes in the behaviour of mammals by Ritzel and Gallo (2020) found alterations in home range and dietary preferences, shifts in activity budgets and vigilance, decreased flight distance and an increase in nocturnal activity in wild mammals living in

a. war zones compared with those in non-war zones

b. forest ecosystems compared with grassland ecosystems

c. urban areas compared with natural ecosystems

d. wealthy residential areas compared with poor residential areas

5.9a Gómez *et al.* (2023) conducted a phylogenetic analysis of same-sex sexual behaviour in mammals and found it to be particularly prevalent in the

a. Artiodactyla

b. Rodentia

c. Perissodactyla

d. Primates

5.10a Individuals of which of the species illustrated in Fig. 5.8 keep in auditory contact using sounds from the creaking and snapping of their foot bones?

a. A

b. B

c. C

d. D

Fig. 5.8.

5.11a Verraux's sifakas (*Propithecus verreauxi*) cool down by

a. panting

b. possessing more sweat glands than other lemurs

    c. hugging trees

    d. lying on cold ground

5.12a. The first observation of a wild Western lowland gorilla (*Gorilla g. gorilla*) using a tool involved an adult female

    a. using a branch to reach fruits suspended beyond her reach

    b. using a rock to break open nuts

    c. using a branch to test the depth of water in a pool

    d. using rocks as weapons by throwing them at individuals in a rival troop

5.13a Many mammals have evolved threat displays that allow two opponents to assess each other's strength without incurring the costs of an all-out fight. Clutton-Brock and Albon (1979) examined encounters between red deer (*Cervus elaphus*) stags (Fig. 9) and observed them going through three stages: roaring, parallel walking, fighting. Of the 50 original encounters observed, only 14 (28%) ended with a fight. This 'game' played by the stags is called

Fig. 5.9.

a. consecutive analysis

b. sequential assessment

c. preventative evaluation

d. successive appraisal

5.14a Laboratory-reared chimpanzees may convey different messages while communicating with other chimpanzees and humans by changing the sequence of the signals used. This is known as changing the

a. syntax

b. lexis

c. context

d. grammar

5.15a In a study of olive baboons (*Papio anubis*) (Fig. 5.10), Packer (1977) noticed that two unrelated males would form coalitions. When they encountered a third male in consort with a female in oestrus one of the two males would sometimes gain access to her and mate. The next time the pair of males encountered a male with a female the other member of the coalition would take the female. This behaviour is known as

Fig. 5.10.

a. eusociality

b. kin selection

c. reciprocal altruism

d. group selection

5.16a **Rauber and Manser (2018) studied sentinel behaviour in meerkats (*Suricata suricatta*) (Fig. 5.11). They found that individuals discriminate between the social information provided by different sentinels and adjust their individual vigilance behaviour depending upon the source of the signals they receive. The extent to which individuals use social information depends upon which characteristic of the signaller?**

Fig. 5.11.

a. age

b. experience

c. dominance status

d. sex

5.17a Phenotypic matching is a mechanism by which an animal may

a. identify safe food organisms even though it has never seen conspecifics eating them

b. determine its position in a dominance hierarchy

c. select environments to which it is phenotypically adapted

d. recognise relatives even if it has never interacted with them

5.18a A lek mating system is used by which of the following species?

a. Blackbuck (*Antilope cervicapra*)

b. Green monkey (*Chlorocebus sabaeus*)

c. Yellow mongoose (*Cynictis penicillata*)

d. Ethiopian wolf (*Canis simensis*)

5.19a A sociogram is a type of network diagram that shows

a. the hierarchy within a group of social animals

b. the degree of association among dyads of animals in a group

c. the outcome of aggressive encounters between individuals

d. the genetic relationship between parents and offspring in a social group

5.20a Male rock hyraxes (*Procavia capensis*) advertise their quality to prospective mates by

a. displaying rhythmic bobbing movements of the head

b. performing a complex dance

c. constructing an elaborate nest

d. producing a rhythmic song

6 Ecology and Genetics

This chapter contains questions concerned with the ecology of mammals – including their population dynamics, feeding relationships and role in modifying habitats – and their genetics.

Foundation

6.1f **Lions, orcas and wolves are all**

    a. vertex predators

    b. summit predators

    c. apex predators

    d. zenith predators

6.2f **A species that engages in ophiophagy feeds on**

    a. fishes

    b. snakes

    c. frogs

    d. earthworms

6.3f **Which of the animals shown in Fig. 6.1 may be described as a riparian mammal?**

    a. A

    b. B

© Paul A. Rees 2024. *Key Questions in Mammalogy* (P.A. Rees)
DOI: 10.1079/9781800624535.0006

c. C

d. D

Fig. 6.1.

6.4f   The chromosomes of a male bonobo (*Pan paniscus*) are denoted by the letters

a. XX

b. XY

c. ZZ

d. ZW

6.5f Approximately what proportion of all of the world's mammal species live in forests?

    a. 32%

    b. 47%

    c. 55%

    d. 68%

6.6f In sheep, black wool is a recessive trait. If the black-wooled phenotype has the genotype *ww* and the white-wooled phenotype has the genotypes *WW* or *Ww*, what proportion of the offspring would be black if a black female was mated with a white male whose mother was black?

    a. 100%

    b. 75%

    c. 50%

    d. 25%

6.7f The activities of beavers (*Castor fiber*) alter habitats by creating wetlands that alleviate flooding, restore native woodlands and improve water quality and habitats for a range of species. For these reasons beavers are considered to be

    a. ecosystem restorers

    b. ecosystem engineers

    c. ecosystem renovators

    d. ecosystem modifiers

6.8f The reintroduction of which of the following species into Yellowstone National Park caused a trophic cascade of ecological change including a decline in elk numbers, the recovery of willow and aspen, a reduction in coyote populations resulting in the increase in survival of pronghorn fawns and an increase in food for scavengers such as bears, ravens and eagles?

    a. Beavers (*Castor fiber*)

    b. Bison (*Bison bison*)

    c. Red wolves (*Canis rufus*)

    d. Grey wolves (*Canis lupus*)

6.9f In most mammals the male is the

a. homogametic sex

b. isogametic sex

c. heterogametic sex

d. metagametic sex

6.10f Which of the following species are generalist feeders (Fig. 6.2)?

Fig. 6.2.

a. ii and iii

b. ii and iv

c. i and iii

d. iii and iv

6.11f   In the food chain below (Fig. 6.3) which of the following statements are true?

i.   The aardvark is a myrmecophagous species

ii. The harvester ant is an osteophage

iii. The aardvark is a tertiary producer or secondary consumer

iv. The hyena is a decomposer

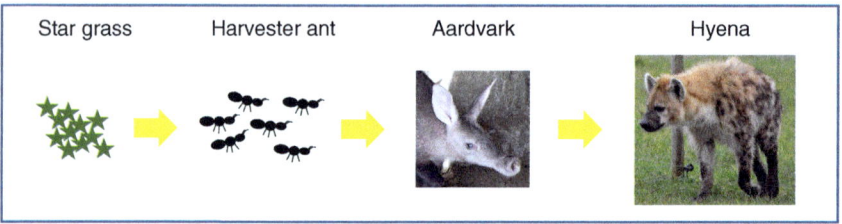

| Star grass | Harvester ant | Aardvark | Hyena |

Fig. 6.3.

   a. i and ii

   b. i and iii

   c. ii and iii

   d. i and iv

6.12f   Which of the following species of African antelope is able to feed on vegetation at the highest level?

   a. Kirk's dik-dik (*Madoqua kirkii*)

   b. Gerenuk (*Litocranius walleri*)

   c. Blue duiker (*Philantomba monticola*)

   d. Water chevrotain (*Hyemoschus aquaticus*)

6.13f   According to Greenspoon *et al.* (2023) the total global biomass of mammals is overwhelmingly dominated by

   a. large wild herbivores and livestock

   b. whales and other marine mammals

   c. large wild herbivores

   d. livestock and humans

6.14f **Some mammals, such as sika deer (*Cervus nippon*) and woodmice (*Apodemus*) possess extra (supernumerary) chromosomes known as**

a. A chromosomes

b. B chromosomes

c. C chromosomes

d. D chromosomes

6.15f **What is the life expectancy of a 15-year-old African elephant (*Loxodonta africana*) if its longevity is assumed to be 70 years?**

a. Seventy years because life expectancy and longevity are the same thing

b. Fifty-five years

c. Up to fifty-five years

d. Eighty-five years

6.16f **Which ecologist is well-known for his use of the historical records of the Hudson's Bay Company to analyse the population dynamics of fur-bearing mammals?**

a. Eugene Odum

b. Robert MacArthur

c. Charles Elton

d. G. Evelyn Hutchinson

6.17f **The number of female live births per female per unit time (usually one year) is known as the**

a. intrinsic rate of natural increase

b. fecundity rate

c. finite rate of increase

d. survivorship rate

6.18f **A leucistic lion (*Panthera leo*) would have fur coloured**

    a. white

    b. yellow

    c. grey

    d. black

6.19f **Some domestic cats have Klinefelter's syndrome. This means their sex chromosomes are**

    a. XYY and present as male

    b. XXX and present as female

    c. XXY and present as female

    d. XXY and present as male

6.20f **A group of individuals of the same age recruited into a population at the same time is known as a**

    a. cohort

    b. consort

    c. cadre

    d. contingent

Intermediate

6.1i **Desert rodents**

    i. are semi-fossorial

    ii. are water-independent

    iii. often plug the entrance to their burrows

    iv. remain in their burrows for long periods of the day

    v. are diurnal

    a. i, iii and iv

    b. i, ii and v

    c. ii, iv and v

    d. i, ii, iii and iv

6.2i Which of the following is least likely to be considered a *K*-selected species?

a. Florida manatee (*Trichechus manatus*)

b. Asian elephant (*Elephas maximus*)

c. Black rat (*Rattus rattus*)

d. North American bison (*Bison bison*)

6.3i Harvesting horns from male saiga antelopes (*Saiga tatarica*) living around the Caspian Sea has resulted in so many males being removed from the population that breeding by females has been compromised. The phenomenon whereby the growth rate of a population begins to decline once the population size falls below a critical low level is known as the

a. Avery effect

b. Adams effect

c. Allen effect

d. Allee effect

6.4i Which of the following traps is not used to catch mammals in the field?

a. Skinner trap

b. Sherman trap

c. Longworth trap

d. Harp trap

6.5i The population of a species of small mammal in a woodland may be calculated using a mark-release-recapture method known as the

a. Clinton Index

b. Nixon Index

c. Truman Index

d. Lincoln Index

6.6i Woolly bats (*Kerivoula*) in Borneo roost in pitcher plants belonging to the genus *Nepenthes*. The bats have a place to hide and rest and in return they provide nutrients to the plants via their faeces. This relationship is known as

a. mutualism

b. biocoenosis

c. comity

d. mutuality

6.7i Some species of pika (*Ochotona*) subsist on a diet consisting largely of

a. nuts

b. insects

c. flowers

d. pine needles

6.8i The populations of African and Asiatic lions are

a. sympatric

b. allopatric

c. allochthonous

d. autochthonous

6.9i The first mammal to be cloned from an adult somatic cell was a

a. goat called Billy

b. rat called Ronald

c. mouse called Minnie

d. sheep called Dolly

6.10i **The red panda (*Ailurus fulgens*) (Fig. 6.4) is active at twilight, that is, it is**

a. crepuscular

b. nocturnal

c. diurnal

d. stochastic

Fig. 6.4.

6.11i **Which of the following species has the lowest intrinsic rate of natural increase (*r*)**

a. Common vole (*Microtus arvalis*)

b. Norway rat (*Rattus norvegicus*)

c. African elephant (*Loxodonta africana*)

d. Wapiti (*Cervus canadensis*)

6.12i **Most mammals have**

a. polygynous mating systems

b. monogamous mating systems

c. polyandrous mating systems

d. polygynandrous mating systems

6.13i **Which of the following is not the name of an interspecies hybrid?**

a. Tigon

b. Liger

c. Horro

d. Zonkey

6.14i **Any of a group of two or more genes that control a continuously variable character – such as milk production in cattle – is known as a**

a. multigene

b. maxigene

c. megagene

d. polygene

6.15i **Which of the survivorship curves shown in Fig. 6.5 is that of a population of humans?**

a. A

b. B

c. C

d. D

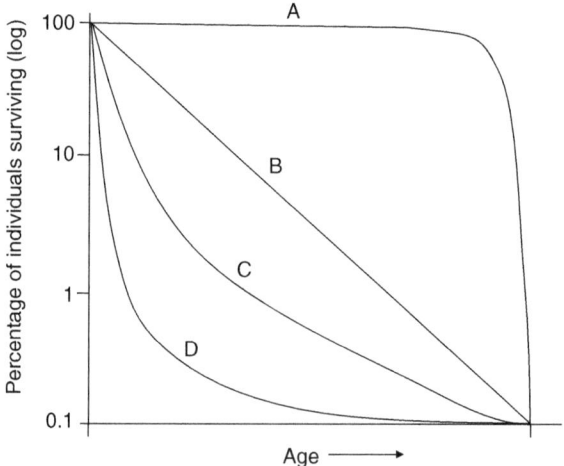

Fig. 6.5.

6.16i **The bats that belong to the Megachiroptera feed primarily on**

a. insects

b. tree sap

c. fruit

d. flowers

6.17i **Which of the following species preys on pacas (*Agouti paca*) (Fig. 6.6 left) and capybaras (*Hydrochoerus hydrochaeris*) (Fig. 6.6 right)?**

Fig. 6.6.

a. Bush dog (*Speothos venaticus*)

b. Swift fox (*Vulpes velox*)

c. Bat-eared fox (*Otocyon megalotis*)

d. Raccoon dog (*Nyctereutes procyonoides*)

6.18i **A recessive gene in some cheetahs (*Acinonyx jubatus*) produces a blotchy coat pattern with black stripes down the spine. This form of the species is known as a**

a. royal cheetah

b. king cheetah

c. queen cheetah

d. pharaoh cheetah

6.19i In the mid-1970s Caughley (1976) hypothesised the exist-
ence of a stable limit cycle to explain the relationship, in
Africa, between

    a. lions and zebras

    b. cheetahs and gazelles

    c. painted dogs and impala

    d. elephants and forests

6.20i Which of the following is not a characteristic of sirenians?

    a. They feed on soft grasses

    b. They are fully aquatic

    c. They have a large caecum that functions as a hindgut fermentation
chamber

    d. They occur exclusively in marine waters

Advanced

6.1a Which of the following represents the sex chromosomes
present in a male duck-billed platypus (*Ornithorhyncus
anatinus*)?

    a. $X_1Y_1 \, X_2Y_2 \, X_3Y_3$

    b. $X_1Y_1 \, X_2Y_2 \, X_3Y_3 \, X_4Y_4$

    c. $X_1Y_1 \, X_2Y_2 \, X_3Y_3 \, X_4Y_4 \, X_5Y_5$

    d. $X_1X_1 \, X_2X_2 \, X_3X_3 \, X_4X_4$

6.2a Which of the following species is important in the dispersal
of the seeds of the cucurbit plant (*Cucumis humifructus*), a
type of cucumber?

    a. African elephant (*Loxodonta africana*)

    b. Aardvark (*Orycteropus afer*)

    c. Waterbuck (*Kobus ellipsiprymnus*)

    d. Geoffroy's horseshoe bat (*Rhinolophus clivosus*)

6.3a **The ages of Dall sheep (*Ovis dalli*) may be estimated post mortem from their skulls by**

a. counting the number of teeth present

b. measuring the size of the foramen magnum

c. measuring the length of the diastema

d. counting the rings on their horns

6.4a **Which of the following populations of 100 individuals has the largest effective population size, assuming each individual is an adult capable of reproduction?**

a. 75 males; 25 males

b. 50 males; 50 females

c. 41 males; 59 females

d. 10 males; 90 females

6.5a **Which of the following statements is not true?**

a. The Y-chromosome in primates is larger than the X-chromosome

b. Ginger coat colour in cats is controlled by a sex-linked recessive allele

c. Cheetah populations in Africa exhibit low genetic diversity as a result of experiencing a genetic bottleneck during their evolutionary history

d. Red-green colour blindness in humans is a sex-linked trait

6.6a **Cooke *et al.* (2019) have suggested that, based on species' extinction probabilities, the non-random nature of the loss of species is likely to result in a shift in ecological strategies – over the next ten years – towards more species that are**

a. large, slow-living, plant-eating, generalists with low fecundity

b. large, fast living, meat-eating, specialists with low fecundity

c. small, fast-living, highly fecund, insect-eating, generalists

d. small, slow-living, plant-eating, specialists with low fecundity

6.7a When a population of voles was enclosed by a fence in an area of pasture the size of a football (soccer) field the population increased in size by approximately five times. The population then crashed due to an increase in competition, aggressive behaviour and a decrease in resources as the animals were unable to disperse from the fenced area. This so called 'fence effect' was first described by

    a. Charles J. Krebs

    b. Eugene P. Odum

    c. G. Evelyn Hutchinson

    d. Frederic Clements

6.8a Which of the following islands was cleared of introduced feral cats that had been destroying its population of burrowing petrels after a control programme that extended from 1977 to 1991?

    a. Bongoyo Island

    b. Marion Island

    c. Davis Island

    d. Friday Island

6.9a A standard technique used to produce an age-specific model of population growth for mammals – and other taxa – is called a

    a. Leslie matrix

    b. Paul matrix

    c. George matrix

    d. John matrix

6.10a Table 6.1 is a hypothetical partial life table for a species of mammal.

Table 6.1. (Adapted from Sinclair et al., 2006).

Age (years) X	f_x	l_x	d_x	q_x
0	1200	1.00	0.58	0.58
1	500	0.42	0.17	0.40
2	300	0.25	0.08	0.32
3	200	0.17	–	–
.	.	.	.	.

Match the abbreviations in Table 6.1 with the correct column headings in Table 6.2

Table 6.2

Abbreviation	A	B	C	D
f_x	Mortality	Survival frequency	Survival frequency	Survival frequency
l_x	Mortality rate	Survivorship	Survivorship	Mortality
d_x	Survival frequency	Mortality rate	Mortality	Survivorship
q_x	Survivorship	Mortality	Mortality rate	Mortality rate

    a. A

    b. B

    c. C

    d. D

6.11a The Transcaucasian mole vole (*Ellobius lutescens*) is unusual because

    a. females possess a Y chromosome

    b. both sexes possess three sex chromosomes

    c. neither sex possesses any sex chromosomes

    d. males do not possess a Y chromosome

6.12a According to Malhi *et al.* (2022), which of the following is least likely to be an effect of the presence of large mammals in temperate, subtropical and tropical grassland ecosystems?

a. An increase in bush and forest fires

b. An increased albedo

c. An increased retention of carbon in the soil and vegetation

d. An increase in habitat heterogeneity

6.13a. The term 'megafauna' is widely used in the study of mammal ecology. According to a review of the use of the term in published work conducted by Moleón *et al.* (2020) much of this literature defines megafauna in relation to the mass of animals and includes species with adults whose mass is

a. 100 to 1000 kg for herbivores and 20 to 150 kg for carnivores

b. greater than 50 kg for herbivores and greater than 25 kg for carnivores

c. greater than 45 to 1000 kg for herbivores and 15 to 100 kg for carnivores

d. up to 1500 kg for herbivores and up to 150 kg for carnivores

6.14a Which of the following is a possible explanation of how the size of populations is regulated in some mammal species whereby an increase in density – and consequent increase in social contact – results in selection for more aggressive individuals with low reproductive rates causing a subsequent reduction in population size?

a. Chitty's hypothesis

b. Southern's hypothesis

c. Kemp's hypothesis

d. Coe's hypothesis

6.15a **Skogland (1985) studied the relationship between reindeer density and the number of calves produced in Norway. Fig. 6.7 suggests that juvenile reindeer mortality is**

   a. density-independent

   b. density-dependent

   c. controlled by stochastic factors

   d. determined by winter temperatures

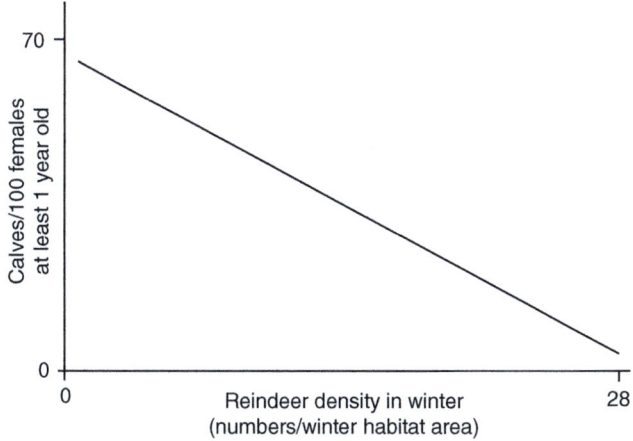

Fig. 6.7.

6.16a **The Lincoln index allows an estimate to be made of the size of a population of small mammals in a particular area by using the following procedure:**

   1. Capture a sample of animals and call the number captured n_1

   2. Mark these individuals and return them to the population

   3. Capture a second sample of animals and call the number captured n_2

   4. Record the number of recaptures (r), i.e. the number of marked animals in the second sample

   5. Calculate an estimate of the size of the population using the Lincoln index using the formula

a. $r / (n_1 \times n_2)$

b. $(n_1 \times r)/n_2$

c. $n_1/(n_2 - r)$

d. $(n_1 \times n_2)/r$

6.17a The Lincoln index assumes that, in the population being sampled, there

a. is no immigration

b. is no emigration

c. are no births or deaths

d. are no birth or deaths and no migration

6.18a According to Greenspoon *et al.* (2023), which of the following mammal groups has the largest global biomass?

a. Elephants

b. Baleen whales

c. Rodents

d. Dolphins and porpoises

6.19a A set of mathematical models that predict the patterns of animal behaviour that might be favoured by natural selection so that members of a species choose a diet that yields the highest energy gain over time is known as

a. optimal foraging theory

b. maximum sustainable yield

c. optimal feeding efficiency

d. cumulative energy gain

6.20a Abramsky (1981) studied populations of two woodmice species – *Apodemus mystacinus* and *A. sylvaticus* – in Israel. The relationship between the two species shown in Fig. 6.8 indicates that

a. intraspecific completion occurs within them

b. interspecific competition occurs between them

c. interspecific competition *may* occur between them

d. their populations are controlled by stochastic processes

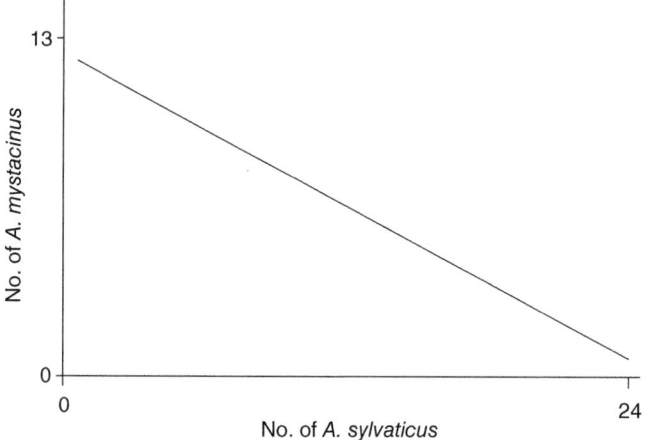

Fig. 6.8.

7 Zoogeography

This chapter contains questions concerned with the current and past geographical distributions of mammals and the factors that have affected their dispersal over geological time.

Foundation

7.1f   Wallace's line is an imaginary boundary which separates the

    a. Oriental and Australasian faunal regions

    b. Palaearctic and Neotropical faunal regions

    c. Ethiopian and Palaearctic faunal regions

    d. Palaearctic and Oriental faunal regions

7.2f   Which of the locations shown in Fig. 7.1 does not contain any extant ursids?

    a. A

    b. B

    c. C

    d. D

© Paul A. Rees 2024. *Key Questions in Mammalogy* (P.A. Rees)
DOI: 10.1079/9781800624535.0007

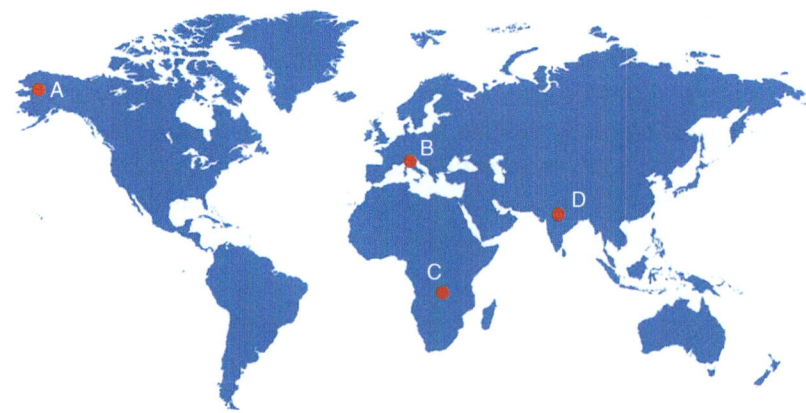

Fig. 7.1.

7.3f   The dhole (*Cuon alpinus*) is a species of canid native to

    a. southern South America

    b. Central America

    c. Sub-Saharan Africa

    d. Asia

7.4f   Who first used the term 'biodiversity hotspot'?

    a. Alfred Russel Wallace

    b. Norman Myers

    c. Edward O. Wilson

    d. Daniel H. Janzen

7.5f   Which of the following species is not an Old World monkey?

    a. Brown capuchin (*Cebus apella*)

    b. Mandrill (*Mandrillus sphinx*)

    c. Black-footed grey langur (*Semnopithecus hypoleucos*)

    d. King colobus (*Colobus polykomos*)

7.6f The land bridge that existed between the Alaskan Peninsula and the Siberian part of Asia at various times was used by some rodent groups, cervids and carnivores to cross from Asia to North America and by camelids to cross in the reverse direction. Only mammals that were adapted to a cold climate could use this bridge. A dispersal route such as this, that only allows certain types of animals to pass through, is known as a

    a. sweepstakes route

    b. filter route

    c. restrictive route

    d. selective route

7.7f In which of the following areas in Fig. 7.2 (A – D) do the natural distributions of extant marsupials and terrestrial placentals overlap?

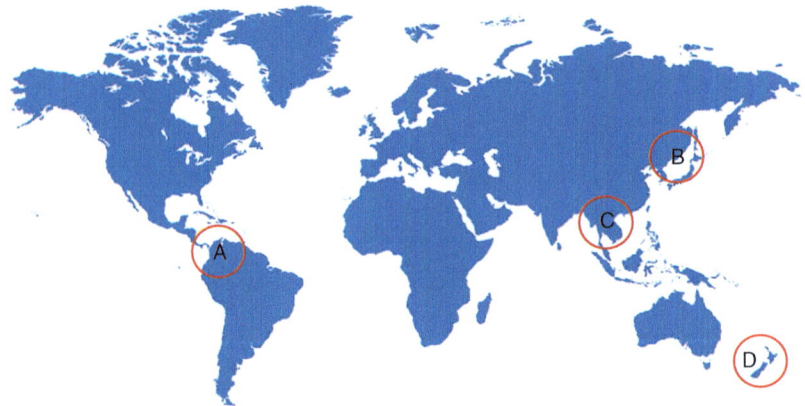

Fig. 7.2.

    a. A

    b. B

    c. C

    d. D

7.8f The greatest numbers of wild bovids and the largest number of bovid species are found in

    a. Africa

    b. Southeast Asia

c. Europe

d. North America

7.9f   Which of the following has been most important in allowing mammals to colonise and remain active in a wide range of habitats across the Earth?

a. The ability to hibernate

b. The possession of a complex brain

c. Variations in dentition

d. Being endothermic

7.10f Which of the following countries has the largest number of mammal species?

a. Indonesia

b. Australia

c. Brazil

d. Madagascar

7.11f Which of the following canids is able to survive at the highest altitude?

a. Bat-eared fox (*Otocyon megalotis*)

b. Tibetan fox (*Vulpes ferrilata*)

c. Fennec fox (*Vulpes zerda*)

g. Golden jackal (*Canis aureus*)

7.12f A population of Barbary macaques (*Macaca sylvanus*) (Fig. 7.3) lives on Gibraltar. These animals were introduced to this area and are native to

a. the Western Ghats

b. the Pyrenees

c. the Atlas Mountains

d. the Ural Mountains

Fig. 7.3.

7.13f Geographical variation in body size often occurs in mammalian species such that individuals tend to be larger in regions that experience lower mean annual temperature and smaller where temperatures are higher. This pattern of variation is known as

    a. a meme

    b. an ecotone

    c. a clone

    d. a cline

7.14f The naked mole-rat (*Heterocephalus glaber*) occurs naturally in which of the locations indicated in Fig. 7.4?

    a. A

    b. B

    c. C

    d. D

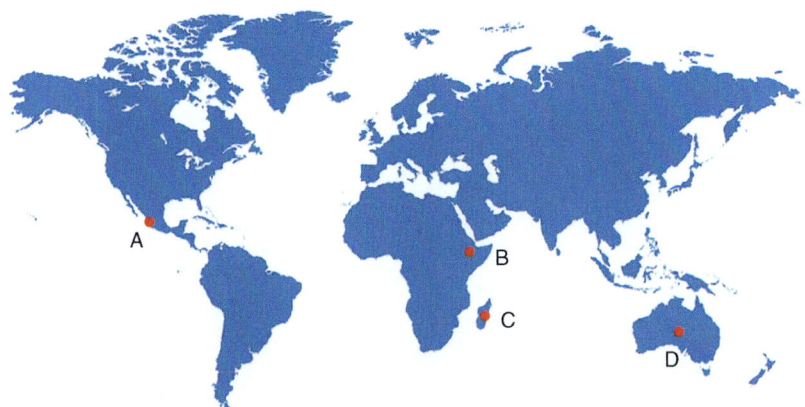

Fig. 7.4.

7.15f The 'island rule' or 'Foster's rule' states that when species migrate to islands they evolve so that

a. large species tend to become smaller than mainland populations

b. small species tend to become larger than mainland populations

c. large species tend to become smaller than mainland populations and small species tend to become larger than mainland populations

d. they tend to have fewer offspring than the mainland populations

7.16f The distribution of the species shown in Fig. 7.5 is confined to

a. Cuba

b. Madagascar

c. Sri Lanka

d. Borneo

Fig. 7.5.

7.17f Differences in the appearance of the Arctic fox (*Vulpes lagopus*) and the fennec (*V. zerda*) provide an illustration of

    a. Allen's rule

    b. Andrew's rule

    c. Archer's rule

    d. Albert's rule

7.18f Which continent has not experienced a mass extinction of its mammalian megafauna?

    a. North America

    b. South America

    c. Europe

    d. Africa

7.19f Which of the following regions has the lowest proportion of mammalian endemics?

    a. Australia

    b. North America

    c. southern Africa

    d. South America

7.20f Apart from bats and rodents (and excluding dingoes), how many native terrestrial eutherian (placental) mammals are found in Australia?

    a. 0

    b. 1

    c. 3

    d. 5

Intermediate

7.1i To which zoogeographical region does the species shown in Fig. 7.6 belong?

Fig. 7.6.

    a. Ethiopian

    b. Oriental

    c. Neotropical

    d. Palearctic

7.2i Which of the following species naturally occurs in the biome known as 'taiga'?

a. Moose (*Alces alces*)

b. Aardvark (*Orycteropus afer*)

c. Klipspringer (*Oreotragus oreotragus*)

d. Guanaco (*Lama guanicoe*)

7.3i Which of the following canids is found in the area known as the Pantanal?

a. Coyote (*Canis latrans*)

b. Maned wolf (*Chrysocyon brachyurus*)

c. Simian wolf (*Canis simensis*)

d. Black-backed jackal (*Lupulella mesomelas*)

7.4i Which of the animals illustrated in Fig. 7.7 belongs to a Nearctic species?

a. A

b. B

c. C

d. D

Fig. 7.7.

7.5i   Which of the following are not native species in the British Isles?

   i.   European rabbit (*Oryctolagus cuniculus*)

   ii. European wildcat (*Felis silvestris*)

   iii. Sika deer (*Cervus nippon*)

   iv. Coypu (*Myocastor coypus*)

   v.   Fallow deer (*Dama dama*)

   a. ii and iv only

   b. i, ii, iii and iv only

   c. ii, iii and iv only

   d. i, iii, iv and v

7.6i   **Only the Ethiopian and Oriental regions contain**

    a. apes, elephants and rhinoceroses

    b. tapirs, elephants and apes

    c. elephants, rhinoceroses and lemurs

    d. apes, bison and tapirs

7.7i   **Which of the following areas has an exceptionally high density of large mammals?**

    a. Ngorongoro Crater, Tanzania

    b. Mole National Park, Ghana

    c. Tsavo East National Park, Kenya

    d. Park National du W, Niger

7.8i   **Which of the following species does not occur in the wild north of Wallace's line?**

    a. Tiger (*Panthera tigris*)

    b. Numbat (*Myrmecobius fasciatus*)

    c. Bornean orangutan (*Pongo pygmaeus*)

    d. Binturong (*Arctictis binturong*)

7.9i   **The extant members of the Xenarthra are found exclusively in**

    a. Australasia

    b. Southeast Asia

    c. the Americas

    d. sub-Saharan Africa

7.10i **The distribution of the species shown in Fig. 7.8 may be described as**

    a. Nearctic

    b. Palaearctic

    c. Oriental

    d. Neotropical

Fig. 7.8.

7.11i Golden moles (Chrysochloridae) are native to

    a. sub-Saharan Africa

    b. South America

    c. New Guinea

    d. Sri Lanka

7.12i Which of the following areas of water is not inhabited by a species of sirenian?

    a. The Caribbean

    b. West coast of Africa

    c. The Mediterranean

    d. The Amazon River Basin

7.13i According to Lehman and Fleagle (2006), primate species diversity is highest in

    a. Africa

    b. Central America

    c. Indonesia

    d. the Neotropics and Asia

7.14i **Which of the following primates is not found in Madagascar?**

    a. Indri (*Indri indri*)

    b. Silky sifaka (*Propithecus candidus*)

    c. Bosman's potto (*Perodicticus potto*)

    d. Aye-aye (*Daubentonia madagascariensis*)

7.15i **New Zealand's only native mammals are**

    a. bats, pinnipeds and cetaceans

    b. bats, cervids and rodents

    c. bats, pinnipeds and cervids

    d. bats, rodents and cetaceans

7.16i **South America has no native species of**

    a. marsupials

    b. ursids

    c. suids

    d. bovids

7.17i **The similarities between the mammalian faunas of North America and Europe are the result of the splitting of the northern land mass known as Laurasia by the process of**

    a. continental drift

    b. land mass slide

    c. geological shift

    d. plate drift

7.18i **The three countries which, taken together, contain the largest number of mammal species are**

    a. Tanzania, India and Australia

    b. China, Mexico and Peru

    c. Indonesia, Brazil and China

    d. Peru, Brazil and the Democratic Republic of Congo

7.19i **Kinkajous (*Potos flavus*) are native to**

a. Central and South America

b. Southeast Asia

c. Madagascar

d. Sri Lanka

7.20i **Which of the following lists of taxa is representative of those that occur in the neotropical region?**

a. Colobus monkey, oribi, leopard, meerkat, cane rat

b. Moose, black bear, beaver, red fox, caribou

c. Sloth bear, chital deer, fishing cat, nilgai, brown palm civet

d. Llama, spider monkey, jaguar, maned wolf, tapir

Advanced

7.1a **Wallace's line passes between**

a. Sumatra and Java

b. Australia and Timor

c. Sumatra and Borneo

d. Borneo and Sulawesi

7.2a **Apart from Madagascar, lemurs are also found in the**

a. Comoro Islands

b. Seychelles

c. Andaman Islands

d. Nicobar Islands

7.3a **Which of the following is not part of the current natural range of the duck-billed platypus (*Ornithorhyncus anatinus*)?**

a. Tasmania

b. Queensland

c. Western Australia

d. New South Wales

7.4a **Which of the following areas of the United States has the highest mammal species richness?**

a. South-west Florida

b. Eastern Texas

c. South-east Arizona

d. Northern Illinois

7.5a **The puma (*Puma concolor*) has a range that extends from**

a. Idaho to Northern Argentina

b. Canada to the southern tip of Chile

c. Wyoming to Paraguay

d. northern Oregon to southern Bolivia

7.6a **Tarsiers are only found on the islands of**

a. the Caribbean

b. New Caledonia

c. Mauritius

d. Southeast Asia

7.7a **Which of the following species are endemic to Madagascar?**

i.   Fossa (*Cryptoprocta ferox*)

ii.   Northern shrew tenrec (*Microgale jobihely*)

iii. Gregarious short-tailed rat (*Brachyuromys ramirohitra*)

iv. Verraux's sifaka (*Propithecus verreauxi*)

v.   Mongoose lemur (*Eulemur mongoz*)

vi. Greater cane rat (*Thryonomys swinderianus*)

a. i, iv and v

b. i, ii, iv, v and vi

c. iii, iv and v

d. i, ii, iii, iv and v

7.8a According to Keast (1969) the Neotropical faunal region owes its numerical richness in mammal species to its large faunas of

    a. rodents and bats

    b. artiodactyls and primates

    c. carnivorans and rodents

    d. primates and bats

7.9a Hot spots of mammalian species richness were identified in two primary regions by Ceballos and Ehrlich (2006). They were:

    a. western North America AND Central and East Africa

    b. northern Mediterranean AND northern South America

    c. Central America and northern South America AND Equatorial Africa (especially East Africa)

    d. Southeast Asia AND Central and South America

7.10a Fig. 7.9 is a photograph of a naturalist who, among other things, was the first to notice that large Amazonian rivers act as barriers to dispersal in primates in what is otherwise a featureless landscape. His name was

Fig. 7.9.

a. Charles Darwin

b. Alfred Russel Wallace

c. Stamford Raffles

d. Charles Lyell

7.11a Increased pigmentation is found in individuals of some mammalian species (e.g. didelphid marsupials) from environments that are cold and dry (lighter colour) to those that are warm and wet (darker colour). This is known as

a. Gloger's rule

b. Allen's rule

c. Jordan's rule

d. Emery's rule

7.12a Which of the following zoogeographical regions has no endemic families of mammals?

a. Neotropical

b. Ethiopian

c. Australian

d. Palaearctic

7.13a Who wrote *Zoogeography: The Geographical Distribution of Animals* in 1957?

a. Philip J. Darlington Jr

b. Alfred Russel Wallace

c. Philip Sclater

d. Ludwig K. Schmarda

7.14a The Great American Biotic Interchange was an event in the late Cenozoic when mammals and other fauna migrated in both directions between North and South America via the

a. Tierra del Fuego Archipelago

b. Caribbean Islands

c. Isthmus of Panama

d. Gulf of California

7.15a The single best predictor of species diversity for bats is

    a. longitude

    b. latitude

    c. altitude

    d. temperature

7.16a Which of the following statements is false?

    a. Some mammals living on islands are smaller than their mainland counterparts.

    b. Many rodents and marsupials living on islands are considerably larger than their mainland counterparts

    c. The fossil elephant form that once inhabited Sicily (*Palaeoloxodon falconeri*) was much smaller than its mainland relatives

    d. The niches of island-dwelling mammals are generally small compared with those of their relatives on the mainland because there is usually less interspecific competition on islands

7.17a In his book *The Mammalian Radiations* Eisenberg (1981) identified eight substrate preferences exhibited by mammals. One of these he termed 'volant' which refers to an animal's ability to

    a. climb

    b. dig

    c. swim

    d. fly

7.18a The process by which the range of a species is divided to form two or more unconnected populations – possibly as a result of the formation of geographical barriers – is known as

    a. covariance

    b. vicariance

    c. dissociation

    d. cleavage

7.19a **The Aplodontiidae (order Rodentia) and Antilocapridae (order Artiodactyla) are endemic to the**

a. western Nearctic region

b. eastern Palearctic region

c. Neotropical region

d. Oriental region

7.20a **Refugia that occur within ice sheets during periods of glaciation allowing fauna and flora to survive the advancing glaciers are known as**

a. nunataks

b. drumlins

c. eskers

d. kames

8 Conservation and Management

This chapter contains questions concerned with the management of populations of wild mammals and the conservation of threatened mammals.

Foundation

8.1f The European badger (*Meles meles*), red fox (*Vulpes vulpes*) and red kangaroo (*Osphranter rufus*) are common species and as such are classified by the International Union for Conservation of Nature (IUCN) as

    a. Little Concern

    b. Least Concern

    c. Less Concern

    d. Minimal Concern

8.2f Attempts have been made to breed back the quagga, an extinct form of

    a. equid

    b. primate

    c. bovid

    d. ursid

8.3f The total number of extant mammal species recognised by the International Union for Conservation of Nature (IUCN) in 2022 was

a. 3,947

b. 6,577

c. 7,190

d. 8,230

8.4f In 1979 the mountain gorilla (*Gorilla beringei beringei*) population was estimated to be fewer than 250. By 2022 the population had

a. fallen by 50%

b. increased by 15%

c. stabilised at around 300

d. increased to over 1,000

8.5f The International Union for Conservation of Nature (IUCN) maintains a list of threatened species – including mammalian species – known as the

a. Green List

b. Red List

c. Rare List

d. Threatened List

8.6f Which of the following mammals is featured in the logo of the World Wildlife Fund?

a. Asian elephant (*Elephas maximus*)

b. Mountain gorilla (*Gorilla beringei beringei*)

c. Giant panda (*Ailuropoda melanoleuca*)

d. Sumatran tiger (*Panthera tigris sondaica*)

8.7f **In which of the following National Parks would you be likely to see a greater one-horned rhinoceros (*Rhinoceros unicornis*) (Fig. 8.1)?**

a. Bandipur National Park

b. Kaziranga National Park

c. Yala National Park

d. Minneriya National Park

Fig. 8.1.

8.8f **Panthera is a conservation organisation whose activities focus on**

a. wild cat species and their habitats

b. species belonging to the genus *Panthera* and their habitats

c. big cats and their habitats

d. lions and their habitats

8.9f **Which of the following mustelid species has been saved from extinction by captive breeding and reintroduction into its natural habitat?**

    a. Honey badger (*Mellivora capensis*)

    b. Black-footed ferret (*Mustela nigripes*)

    c. Japanese marten (*Martes melampus*)

    d. Patagonian weasel (*Lyncodon patagonicus*)

8.10f **Match the species with the correct wild population size in Table 8.1.**

Table 8.1

Species	Wild population size 2023			
	A	B	C	D
Black rhinoceros (*Diceros bicornis*)	6,487	16,803	76	6,487
Javan rhinoceros (*Rhinoceros sondaicus*)	76	4,014	4,014	4,014
Greater one-horned rhinoceros (*Rhinoceros unicornis*)	4,014	76	6,487	76
White rhinoceros (*Ceratotherium simum*)	16, 803	6,487	16,803	16,803

    a. A

    b. B

    c. C

    d. D

8.11f **Very large mammals have outgrown all present-day predators so their populations are most often regulated by**

    a. water supply

    b. food supply

    c. habitat availability

    d. interspecific competition

8.12f Managers of conservation (captive) breeding programmes must take great care to avoid breeding from closely related individuals because of the risk of

a. outbreeding depression

b. inbreeding depression

c. genetic drift

d. heterosis

8.13f The IUCN maintains a list of rare and endangered species, including mammals. Which is the correct sequence of the categories into which such species are placed (Table 8.2)?

Table 8.2

	A	B	C	D
	Extinct	Extinct	Extinct	Extinct
	Extinct in the Wild	Extinct in the Wild	Extinct in the Wild	Extinct in the Wild
	Vulnerable	Critically Endangered	Critically Endangered	Critically Endangered
Increasing likelihood of extinction	Critically Endangered	Vulnerable	Endangered	Endangered
	Endangered	Endangered	Vulnerable	Vulnerable
	Near Threatened	Near Threatened	Near Threatened	Least Concern
	Least Concern	Least Concern	Least Concern	Near threatened
Inadequate data	Data Deficient	Data Deficient	Data Deficient	Data deficient

a. A

b. B

c. C

d. D

8.14f **Which of the following threats to the future survival of polar bears (*Ursus maritimus*) were identified by the Circumpolar Action Plan adopted by the Parties to the Polar Bear Agreement in 2015**

i.   climate change

ii. mineral and energy resource exploitation

iii. pollution

iv. shipping

v.   tourism

vi. parasitism and disease

a. i, ii and iii

b. i, ii, iv and vi

c. i, iii, iv, v and vi

d. All of the above

8.15f **Which of the following areas is protected as a World Heritage Site and is home to river dolphins?**

a. Sundarbans National Park

b. Nahanni National Park

c. Okavango Delta

d. La Amistad International Park

8.16f **Small cetaceans are frequently accidentally captured and killed in fishing nets. This accidental take of animals is known as**

a. incidental catch

b. by-catch

c. subsidiary catch

d. secondary catch

8.17f **In which of the following National Parks would you be likely to see red lechwe (*Kobus leche*)?**

a. Kaziranga National Park

b. Corbett National Park

c. Chobe National Park

d. Korup National Park

8.18f **In 1970-71 around 6,000 elephants died in a drought in which of the following National Parks?**

a. Yankari National Park, Nigeria

b. Tsavo National Park, Kenya

c. Lake Manyara National Park, Tanzania

d. Kruger National Park, South Africa

8.19f **In which of the following states is the Confederated Salish and Kootenai Tribes Bison Range located?**

a. Montana

b. Wyoming

c. Colorado

d. North Dakota

8.20f **Which of the following diseases was deliberately introduced into Australia in 1950 to control the introduced rabbit population?**

a. Rabbit haemorrhagic disease

b. Pododermatitis

c. Babesiosis

d. Myxomatosis

Intermediate

8.1i   Which of the following locations is an important stronghold of the European bison or wisent (*Bison bonasus*) (Fig. 8.2)?

Fig. 8.2.

    a. Bavarian Forest

    b. Białowieża Forest

    c. Hroboskalsko Forest

    d. Letea Forest

8.2i   A moratorium on commercial whaling was instituted in the

    a. 1960s

    b. 1970s

    c. 1980s

    d. 1990s

8.3i   Which of the following protected areas is fenced to prevent poaching of large mammals and reduce human–animal conflict?

    a. Kruger National Park, South Africa

    b. Amboseli National Park, Kenya

    c. Yankari Game Reserve, Nigeria

    d. Tarangire National Park, Tanzania

8.4i 'Operation Oryx' was a conservation project established in 1961 to establish a captive herd of which of the following species aimed at preventing its extinctions?

    a. Scimitar-horned oryx (*Oryx dammah*)

    b. Arabian oryx (*Oryx leucoryx*)

    c. East African oryx (*Oryx beisa*)

    d. Gemsbok (*Oryx gazella*)

8.5i Which of the large mammals shown in Fig. 8.3 is protected by an international treaty signed in 1973 by Canada, Denmark, Norway, the United States and the Union of Soviet Socialist Republics (USSR)?

Fig. 8.3.

a. A

b. B

c. C

d. D

8.6i **Which of the following mammals has been released into Scotland after an absence of over 400 years?**

a. Scottish wildcat (*Felis silvestris*)

b. Pine marten (*Martes martes*)

c. European beaver (*Castor fiber*)

d. Arctic fox (*Vulpes lagopus*)

8.7i **The breeding centre at Chengdu is dedicated to the captive breeding of**

a. Asian elephants (*Elephas maximus*)

b. tigers (*Panthera tigris*)

c. red pandas (*Ailurus fulgens*)

d. giant pandas (*Ailuropoda melanoleuca*)

8.8i **The vaquita (*Phocoena sinus*) is a critically endangered species of**

a. porpoise

b. antelope

c. monkey

d. rodent

8.9i **The last quagga (*Equus quagga quagga*) died in 1883 in**

a. London Zoo

b. Copenhagen Zoo

c. Berlin Zoo

d. Artis (Amsterdam) Zoo

8.10i A fourth echidna species, Attenborough's long-beaked echidna (*Zaglossus attenboroughi*), thought to be extinct and previously only known from a single preserved specimen collected in the 1960s, was rediscovered by an expedition led by scientists from Oxford University in 2023 in

a. Indonesia

b. Australia

c. New Zealand

d. New Guinea

8.11i The Caribbean monk seal (*Neomonachus tropicalis*) was declared extinct by the IUCN in 1994 as a result of

a. loss of breeding sites due to climate change

b. deaths caused by entanglement in fishing nets

c. human trade in its pelts and fat

d. pollution in the Caribbean Sea

8.12i Some large mammals with long generation times may survive loss of habitat and other environmental changes for a long period after they occur before they become extinct. This period between the cause of their extinction and the time when extinction occurs is called the

a. extinction postponement

b. extinction delay

c. extinction deferment

d. extinction debt

8.13i The use of surrogate species is now well established in conservation biology. In 1983 Dr Betsy Dresser supervised the transfer of embryos from a bongo (*Tragelaphus eurycerus*) at Los Angeles Zoo to an individual at Cincinnati Zoo from which of the species shown in Fig. 8.4?

Fig. 8.4.

    a. A

    b. B

    c. C

    d. D

8.14i **Which of the putative subspecies of tiger recognised by Wilting *et al*. (2015) (Table 8.3) were considered to be extinct by the International Union for Conservation of Nature (IUCN) at the end of 2023?**

Table 8.3. Putative subspecies of tiger (*Panthera tigris*). Based on Wilting *et al*. (2015)

Vernacular name	Scientific name
Bengal tiger	*Panthera tigris tigris*
Caspian tiger	*P. t. virgata*
Amur tiger	*P. t. altaica*
Javan tiger	*P. t. sondaica*
South Chinese tiger	*P. t. amoyensis*
Balinese tiger	*P. t. balica*
Sumatran tiger	*P. t. sumatrae*
Indochinese tiger	*P. t. corbetti*
Malayan tiger	*P. t. jacksoni*

    a. Caspian and Malayan

    b. Javan, Balinese and Indochinese

    c. Caspian, Javan and Balinese

    d. Malayan and Indochinese

8.15i **Match the recently extinct mammals shown in Table 8.4 with the locations shown in Fig. 8.5 where they were previously found.**

Table 8.4

Species	A	B	C	D
Giant fossa	3	2	3	4
Bluebuck	1	3	2	1
Lesser bilby	4	4	4	2
Nelson's rice rat	2	1	1	3

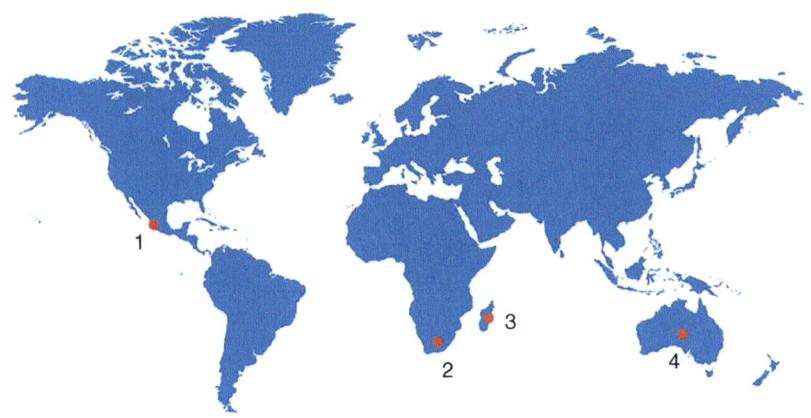

Fig. 8.5.

a. A

b. B

c. C

d. D

8.16i The survival of some rare mammal species is being threatened because they interbreed with related species forming hybrids. For example, feral domestic cats interbreed with wild cats (*Felis silvestris*) in Scotland. This unintended mixing of gene pools is known as

a. interdiction

b. intermission

c. intromission

d. introgression

8.17i The first gorilla born in captivity was born in

a. London Zoo, England in 1960

b. Berlin Zoo, East Germany in 1948

c. Basel Zoo, Switzerland in 1959

d. Columbus Zoo, Ohio, United States in 1956

8.18i Which of the following statements about the use of para-sites for controlling mammal species is true?

    i.   They have never been used

    ii. They have been used extremely successfully in many species

    iii. They have been used and almost universally failed

    iv. The Myxoma virus successfully controlled European rabbits in Australia

    a. i

    b. ii

    c. iii and iv

    d. iii

8.19i According to their 2023 report, the wild mammals most commonly killed by the Wildlife Services operations of the United States Department of Agriculture in 2023 were

    a. badgers

    b. coyotes

    c. grey wolves

    d. foxes

8.20i Sometimes a mammal species becomes extinct because its population has become very small and a chance event wipes it out. This is called

    a. a stochastic extinction

    b. a fatalistic extinction

    c. a driven extinction

    d. a deterministic extinction

Advanced

8.1a **A hypothetical proposal to introduce analogue species of mammals to the Great Plains of the United States – such as elephants in place of mammoths – by Donlan *et al.* (2006) has been referred to as**

a. Pleistocene rewilding

b. Jurassic de-extinction

c. Triassic resurrection

d. Miocene restoration

8.2a **Which mammals were controversially reintroduced into Yellowstone National Park and parts of Idaho by the United States Fish and Wildlife Service (USFWS) in 1995 and 1996?**

a. Grey wolves (*Canis lupus*)

b. Black bears (*Ursus americanus*)

c. Grizzly bears (*Ursus arctos horribilis*)

d. Red wolves (*Canis rufus*)

8.3a **The National Huemul Corridor joins key sites in Chile for fragmented populations of the endangered huemul. This animal is a species of**

a. camelid

b. cervid

c. canid

d. felid

8.4a **The Addo National Park in South Africa is well-known for**

a. the high density of elephants

b. an unusual culture in its chimpanzees

c. its annual wildebeest migration

d. its high density of large predators

8.5a **The first studbook for a wild animal was created for the**

a. giant panda (*Ailuropoda melanoleuca*)

b. okapi (*Okapia johnstoni*)

c. European bison (*Bison bonasus*)

d. Père David's deer (*Elaphurus davidianus*)

8.6a **Central to the efforts to conserve mammal species is the need to determine the number of individuals required in a population to prevent it from becoming extinct. This number is estimated by calculating the**

a. minimum feasible population

b. minimum sustainable population

c. minimum effective population

d. minimum viable population

8.7a **A study published by Vynne *et al.* (2022) demonstrated that the area of the world containing intact large mammal assemblages could be increased by 54 percent – over 11 million km² – by reintroducing or allowing recolonisation by**

a. 20 species

b. 50 species

c. 100 species

d. 200 species

8.8a **The Bramble Cay melomys (*Melomys rubicola*), a small rodent, was declared extinct in 2019 by the Government of Australia, as a result of**

a. hunting

b. disease

c. climate change

d. urbanisation

8.9a **The population of which marine mammal is estimated to have fallen to a minimum of 360 individuals in 1973 from a pre-exploitation size of 239,000? A 1992/1993–2003/2004 circumpolar survey estimated the population to be 2,280 with an estimated annual increase of 8 percent.**

a. Humpback whale (*Megaptera novaeangliae*)

b. Minke whale (*Balaenoptera acutorostrata*)

c. Bowhead whale (*Balaena mysticetus*)

d. Blue whale (*Balaenoptera musculus*)

8.10a **According to Macdonald (2019) the mammal species at greatest risk of extinction are those that fulfil which of the following criteria (Table 8.5)?**

Table 8.5

	A	B	C	D
Home range size	Small	Large	Large	Small
Population density	High	Low	Low	High
Age at weaning	Early	Late	Late	Early
Geographical range	Large	Large	Small	Small
Live where human density is	Low	Low	High	High
Life history	Fast	Fast	Slow	Slow
Dispersal ability	High	High	Low	Low

a. A

b. B

c. C

d. D

8.11a **Which of the following statements about Steller's sea cow (*Hydrodamalis gigas*) is false?**

a. It lived along the coasts of the Komandor Islands

b. It became extinct in 1741

c. Its meat was highly prized

d. This species grew to a length of up to 10 metres

8.12a **Conservation efforts to save rare species from extinction sometimes involve the supplementation of one wild population (A) with individuals from a different population (B). A and B may be different subspecies. For example, American plains buffalo (*Bison bison americanus*) have been used to supplement populations of wood bison (*A. b. athabascae*).**

When subspecies interbreed the new genotypes produced may be inferior to the original stock and less well adapted to the environment occupied. This effect is known as

a. inbreeding depression

b. outbreeding depression

c. heterosis

d. heterozygosis

8.13a **Which disease was used by the federal government of the United States in an attempt to control grey wolf (*Canis lupus*) populations in 1905?**

a. Canine parvovirus

b. Rabies

c. Canine distemper

d. Mange

8.14a **In Los Angeles, California, isolated populations of mountain lions (*Puma concolor*) have been connected by**

a. a vegetated overpass crossing the Ventura Freeway

b. translocating some individuals to other places occupied by mountain lions

c. fifteen newly-created forest corridors

d. three fenced wildlife corridors

8.15a **In southern Africa elephants (*Loxodonta africana*) have been injected with an immunocontraceptive to control their population. This contraceptive is known as PZP and contains proteins obtained from**

a. cattle

b. pigs

c. goats

d. horses

8.16a **In 2024 what percentage of the mammalian species whose conservation status had been assessed by the IUCN were considered to be threatened with extinction?**

    a. 12%

    b. 18%

    c. 26%

    d. 35%

8.17a **Anderson and Scherzinger (1975) demonstrated that un-grazed grassland in the Bridge Creek Wildlife Management Area in Oregon resulted in tall, low quality winter food for elk (*Cervus canadensis*). However, when cattle grazed on the land in spring the grass was kept in a growing state for longer. If cattle were removed before the growing season ended the grasses could regrow significantly to produce a shorter, high quality stand for elk. The introduction of grazing management resulted in elk numbers increasing from 320 to 1,190 between 1964 and 1974. This phenomenon is known as grazing**

    a. facilitation

    b. advancement

    c. assuagement

    d. benefaction

8.18a **Which of the following species have been cloned?**

    i. Black-footed ferret (*Mustela nigripes*)

    ii. Gaur (*Bos gaurus*)

    iii. Crab-eating macaque (*Macaca fascicularis*)

    iv. Mouflon (*Ovis gmelini*)

    v. Banteng (*Bos javanicus*)

    a. i, ii and v

    b. ii, iii and iv

    c. i, ii, iii and v

    d. All of the above

8.19a Animals with long lifespans, large size and overlapping generations that produce few young at a time and prefer stable environments are at higher risk of extinction than those with the opposite characteristics and are referred to as

    a. *r*-strategists

    b. *E*-strategists

    c. *K*-strategists

    d. *t*-strategists

8.20a Jakes *et al*. (2018) recently called for more research to be conducted on the damaging effects of which of the following on wildlife populations?

    a. Roads

    b. Fences

    c. Railway lines

    d. Cities

9 Parasites and Diseases

This chapter contains questions about the parasites and diseases that affect mammals, including humans.

Foundation

9.1f *Taenia solium* is a platyhelminth the adults of which inhabit the

    a. kidneys of mammals

    b. the guts of mammals

    c. the blood vessels of mammals

    d. eyes of mammals

9.2f Rodents may suffer from a bacterial infection known as ulcerative pododermatitis which affects the

    a. eyes

    b. heart

    c. feet

    d. gums

9.3f Which of the following diseases is caused by *Plasmodium coatneyi* in some non-human primate species?

    a. Anaplasmosis

    b. Trypanosomiasis

    c. Dengue

    d. Malaria

© Paul A. Rees 2024. *Key Questions in Mammalogy* (P.A. Rees)
DOI: 10.1079/9781800624535.0009

9.4f *Schistosoma bovis* is a blood fluke that causes intestinal schistosomiasis in

a. ruminants

b. felids

c. rodents

d. marsupials

9.5f Taeniasis is an infection, especially of pigs and cattle, that is caused by various species of

a. roundworm

b. tapeworm

c. protozoans

d. bacteria

9.6f Which of the following taxa can be infected with rabies?

i. Felids

ii. Canids

iii. Ursids

iv. Elephants

v. Rodents

a. i, ii and v

b. i, ii, iii and iv

c. i, ii, iv and v

d. All of the above

9.7f Which disease of mammals is caused by the bacterium *Yersinia pestis*?

a. Plague

b. Q fever

c. Meningitis

d. Botulism

9.8f **Foot-and-mouth disease predominantly affects**

    a. rodents

    b. bats

    c. cloven-hoofed mammals

    d. phocids

9.9f **Which of the animals shown in Fig. 9.1 is most likely to act as the definitive host of the trematode *Fasciola hepatica*?**

Fig. 9.1.

    a. A

    b. B

    c. C

    d. D

9.10f Snow leopards (*Panthera uncia*) are susceptible to multiple ocular colobomas. A coloboma is a condition that affects the

a. eyes

b. ears

c. nose

d. teeth

9.11f *Echinococcus granulosus* is a cestode whose adult form normally lives in the

a. brain of primates

b. liver of felids

c. small intestine of canids

d. large intestine of rodents

9.12f Pernicious anaemia occurs in mammals due to a deficiency of vitamin

a. A

b. K

c. D

d. B_{12}

9.13f Cattle, horses, goats, deer, reindeer and other species may develop marble-sized swellings on their backs caused by the larvae of flies of the genus *Hypoderma*. This condition is known as

a. marbles

b. warbles

c. cobbles

d. hobbles

9.14f Bovine mastitis is an inflammatory response caused by physical trauma or microbial infection of

a. muscle tissue

b. heart tissue

c. lung tissue

d. udder tissue

9.15f   Toxoplasmosis is a disease caused by a

    a. protozoan

    b. cestode

    c. bacterium

    d. prion

9.16f   Feline leukaemia virus (FeLV) affects

    a. felids only

    b. felids and canids

    c. felids, canids and mustelids

    d. felids and a wide range of placental mammals

9.17f   Which of the following statements about obesity in mammals is true?

    a. It may predispose some individuals to develop diabetes

    b. It may make some females of some species infertile

    c. It may cause fatty liver disease in small mammals

    d. All of the above

9.18f   Which of the following is not a cardiac arrhythmia in dogs?

    a. Ventricular premature complex

    b. Atrial fibrillation

    c. Myocarditis

    d. Heart block

9.19f   High levels of faecal cortisol metabolites are indicative of

    a. pregnancy

    b. stress

    c. obesity

    d. respiratory disease

9.20f **Humans with sickle cell anaemia possess red blood cells with an abnormal shape that protects them from infection by the parasite that causes**

    a. sleeping sickness

    b. schistosomiasis

    c. malaria

    d. Chagas disease

Intermediate

9.1i **SFV is an abbreviation for**

    a. simian foamy virus

    b. South Florida virus

    c. severe facial virus

    d. simian facial virus

9.2i **Leptospirosis is a common disease of mammals caused by a**

    a. virus

    b. bacterium

    c. fungus

    d. prion

9.3i **Rinderpest has now been globally eradicated but was a contagious viral disease of**

    a. equids

    b. felids

    c. canids

    d. cloven-hoofed animals

9.4i **Spondylosis is common in old bears and affects the**

    a. jaw

    b. vertebral column

    c. skull

    d. legs

9.5i **Which of the following taxa are not susceptible to monkeypox (M-pox)?**

a. Rodents

b. Anteaters

c. Lagomorphs

d. All of the above are susceptible to this disease

9.6i **Chronic wasting disease (CWD) is caused by a**

a. prion

b. bacterium

c. virus

d. protozoan

9.7i **Orf virus (sore mouth infection) principally infects**

a. monkeys and apes

b. camels and llamas

c. cattle and horses

d. sheep and goats

9.8i **Rift Valley fever is a disease commonly recorded in domesticated mammals in sub-Saharan Africa including sheep, goats, cattle, buffalo and camels. It is caused by a**

a. virus

b. protozoan

c. bacterium

d. prion

9.9i **The disease known as EEHV is a fatal disease that affects**

a. eland

b. echidnas

c. elk

d. elephants

9.10i Which of the following do not suffer from laminitis?

    a. Giraffe

    b. Wallaby

    c. Eland

    d. Takin

9.11i Sarcoptic mange is a skin disease of canids and other mammals caused by

    a. a bacterium

    b. a fungus

    c. a mite

    d. an insect

9.12i Kangaroos are prone to a condition which typically involves swelling around the face and jaw, and infection with *Bacteroides nodosus* and *Fusobacterium necrophorum*. This condition is known as

    a. swollen face

    b. lumpy jaw

    c. swollen jaw

    d. lumpy face

9.13i Bluetongue is a disease that affects

    a. pigs, horses and sheep

    b. dogs, cats and foxes

    c. sheep, camelids, cattle and other ruminants

    d. cattle and other ruminants, horses and other equids

9.14i In order for a country to remain free of the tapeworm *Echinococcus multilocularis* it is important to examine carcasses of

    a. foxes

    b. squirrels

    c. rats

    d. badgers

9.15i   **Which of the following is a notifiable disease of pigs?**

  a. Badowski's disease

  b. Gregorek's disease

  c. Kolodziej's disease

  d. Aujeszky's disease

9.16i   **In captive elephants, a trunk wash is used to diagnose**

  a. metritis

  b. tuberculosis

  c. nephritis

  d. filariasis

9.17i   **Which of the following diseases of seals is caused by a protozoan?**

  a. Phocine distemper

  b. Seal pox

  c. Leptospirosis

  d. Giardiasis

9.18i   **Capture myopathy is a condition that causes damage to the muscles of wild animals. It is**

  a. non-infectious

  b. caused by a viral infection

  c. caused by a bacterial infection

  d. a congenital disease

9.19i   **Which of the following most accurately reflects the number of species of macroparasites that can affect humans?**

  a. Fewer than 50

  b. Fewer than 75

  c. Fewer than 100

  d. More than 150

9.20i **Severe acute respiratory syndrome (SARS) appeared in humans in 2002 probably because people in China kept and lived closely with**

    a. raccoon dogs (*Nyctereutes procyonoides*)

    b. Asian badgers (*Meles leucurus*)

    c. civets (*Viverra zibetha*)

    d. Siberian weasels (*Mustela sibirica*)

Advanced

9.1a **Leishmaniasis is a vector-borne disease of the tropics and subtropics that is transmitted to humans by**

    a. ticks

    b. mosquitoes

    c. sandflies

    d. mites

9.2a **Baylisascaris is a disease of raccoons (*Procyon lotor*) (Fig. 9.2) – caused by a nematode – which may become infected when they eat a paratenic host. This may be a small mammal or bird**

    a. that acts as a reservoir of the parasite but is not necessary to its life cycle

    b. that is essential to the life cycle of the parasite

    c. that acts as an alternative definitive host

    d. in whose body larvae develop into adult worms

Fig. 9.2.

9.3a **Populations of the Tasmanian devil (*Sarcophilus harrisii*) have been decimated by**

    a. a facial cancer

    b. rabies

    c. a parasitic infection

    d. coronavirus

9.4a **Australian bat lyssavirus is related to the virus that causes**

    a. avian influenza

    b. African swine fever

    c. bluetongue

    d. rabies

9.5a **Captive black rhinoceroses (*Diceros bicornis*) suffer from hepatopathy. This is a disease of the**

    a. kidneys

    b. heart

    c. liver

    d. lungs

9.6a **Which of the following is not a common sign of rabies in a mammal?**

a. Facial distortion

b. Excessive drooling

c. Difficulty walking

d. All of the above are common signs of rabies

9.7a **Glanders and farcy are two forms of a disease caused by the bacterium *Burkholderia mallei*. The disease principally affects the skin and respiratory tract of**

a. rats, mice and squirrels

b. dolphins and whales

c. lions, tigers, leopards and jaguars

d. horses, zebras, mules and donkeys

9.8a **Lyme disease is a zoonotic disease that affects humans caused by a spirochaete bacterium. The most important reservoirs of the disease are found in**

a. rodents and deer

b. birds

c. cattle and horses

d. pigs

9.9a **Which of the following are true in relation to cetacean diseases and health?**

i. Wild populations of cetaceans are infected with antibiotic-resistant bacteria

ii. Morbillivirus has been implicated in the most notable die-offs of cetaceans in the Mediterranean Sea, Atlantic Ocean, Indian Ocean and Gulf of Mexico

iii. Marine contaminants bioaccumulate in the blubber of cetaceans

iv. Polychlorinated biphenyls (PCBs) have not been found in whales at concentrations that are high enough to compromise their health

a. i and ii

b. ii and iii

c. i, ii and iii

d. All are true

9.10a **Which of the following animals living in zoos have contracted COVID-19?**

i. Hippopotamus

ii. Tiger

iii. Binturong

iv. Snow leopard

v. Lion

vi. Gorilla

a. ii, iv and v only

b. i, iii and vi only

c. ii iv, v and vi only

d. COVID-19 has been recorded in all of these species

9.11a **Historically, nonhuman primates living in zoos have been provided with food with a high sugar content – principally commercially produced fruits – and have had limited opportunity for exercise. Approximately what percentage of female primates examined in North American zoos in a study published in 2013 (Kuhar *et al.*, 2013) were diagnosed with diabetes?**

a. 20%

b. 40%

c. 60%

d. 80%

9.12a **Which of the following is not a disease of mammals?**

a. Bovine ephemeral fever

b. Sacbrood

c. Rocky mountain spotted fever

d. African horse sickness

9.13a Which of the following mammalian diseases is not tick-borne?

a. Trichinosis

b. Babesiosis

c. Anaplasmosis

d. Tularaemia

9.14a What condition could a pressure plate system be used to diagnose in a rhinoceros living in a zoo?

a. Heart disease

b. Early chronic foot disease

c. Chronic respiratory disease

d. Tuberculosis

9.15a In mammals, an allostatic load index would be used to measure

a. the physical fitness of the body

b. the digestibility of the foods consumed

c. the cumulative burden of chronic stress on the body

d. the load experienced by the legs resulting from the body mass

9.16a A comparative study of the dentition of captive and wild brown bears (*Ursus arctos*) conducted by Wenker *et al.* (1999) found that the likely cause of canine tooth and secondary alveolar lesions in the zoo bears was

a. stereotypic chewing

b. poor diet

c. inadequate cleaning opportunities

d. inappropriate dental interventions

9.17a **A study of mandibular fractures in giraffes (*Giraffa camelopardalis*) (Fig. 9.3) living in European zoos (Remport *et al.*, 2022) found that 50 percent of cases were the result of**

a. congential abnormalities

b. aggressive interactions with conspecifics

c. vitamin D deficiencies

d. poorly-designed hay racks

Fig. 9.3.

9.18a **Which of the following diseases does not affect elephants?**

a. Anthrax

b. Foot and mouth disease

c. Rhabdiasis

d. Schistosomiasis

9.19a **Which of the following is an infection caused by roundworms transmitted in the faeces of cats and dogs that may cause eye damage in humans, particularly children?**

a. Toxocariasis

b. Toxoplasmosis

c. Giardiasis

d. Leishmaniasis

9.20a **Which mammals are thought to be the natural hosts and principal animal reservoir of *Zaire ebolavirus* (EBOV) in Africa by the World Organisation for Animal Health?**

a. Rodents

b. Fruit bats

c. Chimpanzees

d. Baboons

10 Domestication and the Human Use of Mammals

This chapter contains questions about the process of domestication, types of domestic mammals and their role as a source of food and materials, as beasts of burden and companion animals, and their role in research.

Foundation

10.1f Which of the following is not a breed of horse?

    a. Suffolk Punch

    b. Russian Don

    c. Cleveland Bay

    d. Dexter

10.2f Elephants cannot be considered domesticated because

    a. they have not been selectively bred and genetically adapted for human use

    b. they still occur in the wild

    c. they are not kept to produce food, hides or other materials for human use

    d. attempts at breeding them in captivity have been largely unsuccessful

© Paul A. Rees 2024. *Key Questions in Mammalogy* (P.A. Rees)
DOI: 10.1079/9781800624535.0010

10.3f **Which of the following mammals have been used as pack animals?**

i. Yaks

ii. Dogs

iii. Water buffalo

iv. Reindeer

v. Goats

a. i, iii and iv

b. ii, iv and v

c. i, iii, iv and v

d. All of them

10.4f **The Basenji, Azawakh, Beauceron and Chinook are all breeds of domestic**

a. rabbits

b. cats

c. dogs

d. goats

10.5f **Which of the following is not a breed of domestic cattle?**

a. Belgian blue

b. Hinterwald

c. Normande

d. Bagot

10.6f Match the images in Fig. 10.1 with the names of sheep breeds in Table 10.1.

Fig. 10.1.

Table 10.1.

Image	A	B	C	D
1	Texel	Valais blacknose	Manx Loaghtan	Texel
2	Dutch spotted	Manx Loaghtan	Valais blacknose	Valais blacknose
3	Manx Loaghtan	Dutch spotted	Dutch spotted	Dutch spotted
4	Valais blacknose	Texel	Texel	Manx Loaghtan

a. A

b. B

c. C

d. D

10.7f   The horse breed shown in Fig. 10.2 is

   a. a Shire horse

   b. a Flemish horse

   c. a Jutland horse

   d. a Clydesdale horse

Fig. 10.2.

10.8f   Zebu cattle originated in

   a. East Africa

   b. the Indian subcontinent

   c. Southeast Asia

   d. the Middle East

10.9f   The first mammal to make an orbital spaceflight around the Earth was a

   a. cat

   b. dog

   c. monkey

   d. chimpanzee

10.10f Castoreum is a yellow exudate that was once widely used to make perfume. It is produced by

a. beavers

b. otters

c. badgers

d. skunks

10.11f All of the domestic cattle in the world are descended from the

a. chianina

b. auroch

c. pinzgauer

d. salers

10.12f Which of the following species is used in research on leprosy because it is naturally susceptible to, and is a reservoir of, the disease?

a. Duck-billed platypus (*Ornithorhyncus anatinus*)

b. Hoffmann's two-toed sloth (*Choloepus hoffmanni*)

c. Sunda pangolin (*Manis javanica*)

d. Nine-banded armadillo (*Dasypus novemcinctus*)

10.13f Which of the following is not a hybrid between two species of equid?

a. A zonkey

b. A mule

c. A burro

d. A zorse

10.14f Which of the following horse breeds is not classified as a heavy horse?

a. Suffolk Punch

b. Schleswig Coldblood

c. Clydesdale

d. Welsh Cob

10.15f   In medieval England, a warrener raised

    a. goats

    b. rabbits

    c. cattle

    d. sheep

10.16f   The modern horse was domesticated from wild horses by the tribes inhabiting

    a. the prairies of North America

    b. the pampas of South America

    c. the savannah of the Sudan

    d. the Eurasian steppes

10.17f   Approximately how many breeds of domestic cattle exist?

    a. 300

    b. 500

    c. 800

    d. 1,100

10.18f   A genetically-engineered mouse in which a gene has been inactivated in order to study its function is known as a

    a. knock-down mouse

    b. knock-out mouse

    c. knock-off mouse

    d. switch-off mouse

10.19f   Which dog breed was quickly adopted as a carriage dog used to protect carriages from highwaymen after it was introduced into Britain in the eighteenth century?

    a. Dalmatian

    b. Rottweiler

    c. Alsatian

    d. Bullmastiff

10.20f Put the animals in Table 10.2 in the order in which they were domesticated.

Table 10.2

Years BCE	A	B	C	D
34,000-13,000	Dog	Horse	Sheep	Dog
9,000	Cow	Dog	Dog	Sheep
8,300	Horse	Cow	Horse	Cow
3,500	Sheep	Sheep	Cow	Horse

    a. A

    b. B

    c. C

    d. D

Intermediate

10.1i The dromedary (*Camelus dromedarius*) was most likely descended from wild ancestors in

    a. North Africa

    b. the Arabian Peninsula

    c. Egypt

    d. Iran

10.2i The Oncomouse is a genetically-modified laboratory mouse used to study

    a. animal behaviour

    b. depression

    c. cancer

    d. parasitic diseases

10.3i Almost all British deer farms keep

    a. fallow deer (*Dama dama*)

    b. sika deer (*Cervus nippon*)

c. red deer (*Cervus elaphus*)

d. roe deer (*Capreolus capreolus*)

10.4i **Which of the following companies is famous for the fur trading activities it established in Canada in the seventeenth century?**

a. St Lawrence River Company

b. Nelson River Company

c. Cambridge Bay Company

d. Hudson Bay Company

10.5i **Which of the following dog breeds is not considered suitable as a guide dog for a visually-impaired person?**

i. Golden Retriever

ii. Labrador Retriever

iii. Australian Shepherd

iv. Poodle

v. Border Collie

vi. Airedale

a. i, ii and v

b. i, ii and iii

c. i, ii, iii, iv and vi

d. All of these breeds are suitable

10.6i **Which of the following cat breeds is not hairless?**

a. Sphynx

b. Abyssinian

c. Bambino

d. Donskoy

10.7i **Which of the following animals kept as pets originated in Syria?**

    a. golden hamster

    b. common degu

    c. guinea pig

    d. chinchilla

10.8i **Dogs were used in World War I to**

    i. haul machine guns

    ii. carry messages

    iii. locate wounded soldiers

    iv. carry first aid kits

    v. catch rats in the trenches

    a. ii, iv and v

    b. i, iii and v

    c. i, iii, iv and v

    d. All of the above

10.9i **In which of the following countries is cavy farming a well-established and important source of meat for humans?**

    a. Madagascar

    b. Spain

    c. Ecuador

    d. Algeria

10.10i **The opposite of domestication is**

    a. vernalisation

    b. feralisation

    c. socialisation

    d. familiarisation

10.11i Artificial insemination was first achieved in which of the following mammals?

a. Cats

b. Cattle

c. Dogs

d. Pigs

10.12i The meat produced from water buffalo is known as

a. squab

b. chevon

c. venison

d. carabeef

10.13i Semi-nomadic reindeer herding is practised by the

a. Inuit

b. Sámi

c. Bathari

d. Karuk

10.14i Which type of monkey is often trained to work as a helper or service monkey to assist humans?

a. Capuchin

b. Colobus

c. Vervet

d. Red titi

10.15i Domestic dogs are widely kept as companion animals. However, some breeds were originally developed as fighting dogs and, because of the risk they pose to the public, the keeping, breeding and sale of these animals has been banned in some jurisdictions. In Great Britain the relevant legislation is the Dangerous Dogs Act 1991. Which of the following is not a banned breed under this Act?

a. Japanese Tosa

b. Leonberger

c. Fila Brazilieros

d. Dogo Argentinos

10.16i **Which of the following countries have banned experimentation on great apes?**

a. United Kingdom

b. New Zealand

c. Austria

d. All of the above

10.17i **All of the species shown in Fig. 10.3 are South American camelids that are used to produce wool. Match the images with the species listed in Table 10.3**

Fig. 10.3.

Table 10.3

Image number	A	B	C	D
1	Vicuna	Guanaco	Alpaca	Alpaca
2	Alpaca	Llama	Vicuna	Vicuna
3	Llama	Alpaca	Guanaco	Llama
4	Guanaco	Vicuna	Llama	Guanaco

    a. A

    b. B

    c. C

    d. D

10.18i **Which of the following is not a hybrid cat created by breeding a domestic cat with a wild felid species?**

    a. Bengal cat

    b. Himalayan cat

    c. Savannah cat

    d. Cheetoh

10.19i **Which of the following dog breeds is most prone to develop hip dysplasia?**

    a. Great Dane

    b. Whippet

    c. Bull dog

    d. German shepherd

10.20i **The Droughtmaster is a breed of**

    a. cattle from Australia

    b. goat from the Middle East

    c. horse from Mongolia

    d. sheep from northern Niger

Advanced

10.1a In which of the following former colonial territories were African elephants trained as draft animals in the early twentieth century?

a. Southern Rhodesia

b. Belgian Congo

c. Bechuanaland

d. Nyasaland

10.2a Which of the following species has been domesticated and used to detect land mines and tuberculosis using olfaction?

a. Gambian pouched rat (*Cricetomys gambianus*)

b. Lesser cane rat (*Thryonomys gregorianus*)

c. Vangunu giant rat (*Uromys vika*)

d. Manus spiny rat (*Rattus detentus*)

10.3a Pacas (*Cuniculus*) are farmed for their meat in Costa Rica. They are members of the

a. Lagomorpha

b. Perissodactyla

c. Artiodactyla

d. Rodentia

10.4a Which of the following dog breeds is not traditionally used as a livestock guardian dog?

a. Shiba Inu

b. Great Pyrenees

c. Kuvasz

d. Cane Corso

10.5a Almost eight million mummies were found at a catacomb in Saqqara, Egypt in 2013 (Nicholson *et al.*, 2015). They were all mummies of

a. cats

b. dogs

c. humans

d. bulls

10.6a The domestication of mammals has resulted in selection for the retention of juvenile characteristics. This is known as

a. neoteny

b. neophilia

c. neophobia

d. neoplasia

10.7a Who discussed the effects of artificial selection on the characteristics of domesticated mammals and other animals in his book *The Variation of Animals and Plants under Domestication*?

a. Alfred Russel Wallace

b. Charles Darwin

c. Julian Huxley

d. Richard Owen

10.8a A famous long-term study of domestication in a canid species that began in 1959 used

a. raccoon dogs (*Nyctereutes procyonoides*)

b. grey wolves (*Canis lupus*)

c. silver foxes (*Vulpes vulpes*)

d. coyotes (*Canis latrans*)

10.9a The study described in Q10.8a was conducted in

a. Braganca, Portugal

b. Novosibirsk, Siberia

    c. Kristiansand, Oslo

    d. Stargard, Poland

10.10a Which of the following peoples are traditionally yak (*Bos grunniens*) herders?

    a. Brokpa

    b. Lakota

    c. Sámi

    d. Inuit

10.11a When cattle breeders select for high milk yield they are applying

    a. natural selection

    b. stabilising selection

    c. disruptive selection

    d. directional selection

10.12a Lanolin is used in the manufacture of cosmetics and is a wax produced by the sebaceous glands of

    a. beavers

    b. muskrats

    c. sheep

    d. goats

10.13a Which of the following species has been extensively studied by biomedical scientists because individuals have an unusually long life span, show no signs of ageing – such as cancer or neurodegeneration – and exhibit extreme resistance to hypoxia and hypercapnia?

    a. Aardvark (*Orycteropus afer*)

    b. Manatees (*Trichechus* spp.)

    c. Pangolins (*Manis* spp.)

    d. Naked mole-rat (*Heterocephalus glaber*)

10.14a The role of the gut microbiome in the physiology of mammals has been investigated by Manca *et al.* (2020) using animals that lack this microbiome. The animals used in this study were

   a. bacteria-free rats

   b. germ-free mice

   c. virus-free rabbits

   d. germ-free monkeys

10.15a What percentage of the 1,512,210 experimental procedures conducted on animals in Great Britain in 2022 involved monkeys?

   a. 0.15%

   b. 0.52%

   c. 1.73%

   d. 3.62%

10.16a The Draize test was devised in 1944 and was once widely used to test the ocular toxicity of chemicals by dropping small quantities of the test chemical into the eyes of restrained animals. The animals that were most widely used were

   a. dogs

   b. rats

   c. monkeys

   d. rabbits

10.17a Which of the following species are used by the United States Navy's Marine Mammal Program to recover objects in the sea, detect and track undersea targets and detect and apprehend unauthorised swimmers and divers?

   i. Bottlenose dolphin (*Tursiops truncatus*)

   ii. California sea lion (*Zalophus californianus*)

   iii. Harbour seal (*Phoca vitulina*)

   iv. Harbour porpoise (*Phocoena phocoena*)

   v. Short-beaked common dolphin (*Delphinus delphis*)

a. i and ii

b. i and iii

c. i, ii, iii and iv

d. All of the above

10.18a The meat of all of the species listed below is used for human consumption. Which of these species is not farmed?

a. American bison (*Bison bison*)

b. Yak (*Bos grunniens*)

c. Red kangaroo (*Osphranter rufus*)

d. Paca (*Cuniculus paca*)

10.19a Domesticated mammals exhibit distinct changes in their characteristics compared with their wild counterparts. Which of the following changes is not the result of domestication?

i. changes in horn size, shape and growth rate

ii. a greater variety of coat colours and patterns

iii. reduced flight distances

iv. an increase in body size

v. earlier puberty

vi. larger litter sizes

a. ii, iii, iv and vi

b. iii, iv, v and vi

c. iv only

d. i, ii and iii

10.20a Meal bone, oil, spermaceti, ambergris, and extracts from the liver and endocrine glands are all products used by humans and obtained from

a. whales

b. tigers

c. bears

d. rhinoceroses

1. The History and Principles of Mammalogy

1.1f	C	Gerontology is the study of ageing.
1.2f	A	*Walker's Mammals of the World* is a major text written by Ernest P. Walker and later revised by Ronald M. Nowak. Walker was an American biologist who worked for the US Bureau of Fisheries and later became Assistant Director of the National Zoological Park.
1.3f	C	C is correct.
1.4f	C	Mammals possess three bones in the middle ear and a *left* aortic arch.
1.5f	D	The mammalian circulatory system is enclosed within blood vessels and blood circulates alternately between the heart and the lungs and then the heart and the rest of the body.
1.6f	A	*The Naked Ape* was written by Dr Desmond Morris. He was once Curator of Mammals at London Zoo and his book examined the behaviour of humans from the point of view of a zoologist.
1.7f	B	B is correct.
1.8f	D	The Latin word 'mamma' means breast.
1.9f	B	In the nineteenth century they included the physicians John Godman and Harrison Allen, the artist John James Audubon and the clergymen John Bachman.
1.10f	D	Mammalogists study all aspects of the biology of mammals.
1.11f	C	Jean-Baptiste Lamarck was a French zoologist and palaeontologist who believed in the inheritance of acquired characteristics.
1.12f	B	The Etruscan shrew (*Suncus etruscus*) is the world's smallest mammal.

1.13f	C	According to the Mammal Society, 107 species of mammals are found in and around the coast of the British Isles.
1.14f	A	Linnaeus was the first person to use the term 'Mammalia' to describe the mammals.
1.15f	A	Only mammals have a true diaphragm.
1.16f	D	Ernest Neal was a British naturalist and school teacher who spent much of his life studying badgers.
1.17f	B	Baird was, at one time, curator at the Smithsonian Institution.
1.18f	C	John Gould was an English naturalist and taxidermist who worked at the Zoological Society of London. He travelled to Australia, studied and painted its wildlife and published a three-volume work on Australian mammals.
1.19f	B	Raffles worked for the British East India Company, became Governor of Singapore and was a founder of the Zoological Society of London.
1.20f	C	Dr Derek Yalden was an expert on British mammals and worked at the University of Manchester for most of his life.
1.1i	D	Johnston sent the first pieces of hide of an okapi (*Okapia johnstoni*) to the British Museum.
1.2i	A	Schaller published *The Serengeti Lion*, an account of his work on the ecology and behaviour of lions in the Serengeti, Tanzania, in 1972.
1.3i	C	*Mammal Review* is a peer-reviewed academic journal published in the United Kingdom by the Mammal Society.
1.4i	D	D is correct.
1.5i	A	Mammalian hair is made of the protein keratin.
1.6i	B	Coenraad Jacob Temminck was a Dutch zoologist.
1.7i	A	Linnaeus was the first to use the term 'primates'.
1.8i	C	Prof. Tim Clutton-Brock is a Cambridge University mammalogist and behavioural ecologist who undertook long-term studies of red deer (on the Scottish Island of Rhum), meerkats and Soay sheep.
1.9i	D	This is Bermann's rule.
1.10i	A	Frans de Waal is a Dutch-American zoologist who has studied chimpanzees extensively in captivity and wrote *Chimpanzee Politics*.
1.11i	D	J.Z. Young was an English zoologist and neurophysiologist who wrote many books including *The Life of Mammals* (1957) and *The Life of Vertebrates* (1950). David Attenborough later produced a television series for the BBC called *The Life of Mammals*.
1.12i	C	Major texts on the ecology of these large mammals were published in the 1970s.
1.13i	A	This was written by the naturalists Bachman (a clergyman) and Audubon (an artist).

1.14i	C	Mammals were collected by these men as they crossed the continent surveying routes for a railroad.
1.15i	D	Napier was a British surgeon, primatologist and palaeontologist.
1.16i	B	They were all famous palaeontologists and anatomists who made important discoveries relating to the evolution of humans.
1.17i	B	The lesula and kipunji are primates, the khanyou is a rodent, the saola is a bovid.
1.18i	C	Linnaeus named the cougar *Felis concolor*.
1.19i	A	Joseph Grinnell (1877–1939) was the first director of the Museum of Vertebrate Zoology, University of California, Berkeley.
1.20i	A	In 1856 Richard Owen was appointed to the post of Superintendent of the natural history collections of the British Museum.
1.1a	D	Muybridge was an English photographer who had an interest in animal and human motion.
1.2a	B	G.G. Simpson was an American palaeontologist and evolutionary biologist who held posts as Curator at the American Museum of Natural History and at Harvard University's Museum of Comparative Zoology.
1.3a	C	C is correct.
1.4a	B	The pygmy hippopotamus (*Choeropsis liberiensis*) was first reported by Samuel Morton in 1844. The giant panda (*Ailuropoda melanoleuca*) was described by Armand David in 1869. The okapi (*Okapia johnstoni*) was described by Philip Sclater in 1901. The bonobo (*Pan paniscus*) was first recognised as a distinct species in 1928 by Ernst Schwartz.
1.5a	A	A is correct. The Biological Survey advocated killing predators, especially wolves, to protect livestock.
1.6a	A	H. F. Osborn, an American palaeontologist and geologist, wrote this. He was President of the American Museum of Natural History (1908–1935).
1.7a	C	L. Harrison Matthews was a British zoologist. He studied many species from hyenas in Tanzania to the whales of the Southern Ocean.
1.8a	D	Dawson's 'missing link' fossil was a fake made up of the mandible and teeth of an orangutan and a human cranium. This was discovered when the remains were examined by scientists from the Natural History Museum and Oxford University.
1.9a	D	D is correct
1.10a	B	Aristotle used these three divisions.
1.11a	C	Burt made significant contributions to the study of territoriality and home range.
1.12a	B	B is correct.

1.13a	A	Sarah Hrdy is an American anthropologist who has made important contributions to our understanding of the evolutionary basis of female behaviour in primates, including humans.
1.14a	A	A is correct. The neocortex is involved in higher-order brain functions including cognition, language and spatial reasoning.
1.15a	D	In contrast to North America, mammalogists were slow to organise in Europe. The Societas Europea Mammalogica was founded in 1988.
1.16a	B	Australian Mammal Society was founded in 1958.
1.17a	C	The *Atlas of European Mammals* was published in 1999.
1.18a	A	A is correct
1.19a	A	Grinnell was the first director of the Museum of Vertebrate Zoology at the University of California, Berkeley. He devised a detailed protocol for recording observations in the field using a field notebook, a field journal, a species account and a catalogue of specimens.
1.20a	B	This was published by Georges Cuvier and Etienne Geoffrey Saint-Hilaire.

2. Origins, Evolution and Taxonomy

2.1f	B	B is correct.
2.2f	D	D is correct.
2.3f	B	Mammal-like forms first appeared in the fossil record in the late Triassic.
2.4f	D	All of these terms have been used for what are generally termed 'stem-mammals': non-mammalian synapsids.
2.5f	A	'Hippus' means horse in Latin
2.6f	C	This is convergent evolution. The two taxa have evolved in parallel in different parts of the world in response to similar selective pressures, and not from a common ancestor (divergent evolution).
2.7f	B	The Plesiosauria is a clade (order) of extinct Mesozoic long-necked marine reptiles.
2.8f	A	Rhinoceroses belong to the mammalian order Perissodactyla which also contains the tapirs and horses. The species illustrated is a black rhinoceros (*Diceros bicornis*).
2.9f	C	The nominotypical or nominate subspecies is the population from which the species was originally described and its scientific name repeats the name of the species.
2.10f	B	Pygmy possums are marsupials.
2.11f	C	These animals were small, similar to shrews and almost certainly nocturnal.

2.12f	B	*Gigantopithecus* means 'giant ape'.
2.13f	A	Fig. 2.2 is a model of a smilodon. *Smilodon* is an extinct genus of saber-toothed cats from the late Pleistocene.
2.14f	D	Glyptodonts are an extinct clade of large armadillos.
2.15f	D	*Proconsul* was an early ape whose descendants gave rise to the hominoid line.
2.16f	C	The largest extant rodent is the capybara.
2.17f	C	Jonathan Kingdon is a zoologist and artist who has spent much of his life working on the evolution and taxonomic illustration of African mammals.
2.18f	A	These scientific names are synonyms as they all refer to the same species.
2.19f	B	B is the correct chronological order.
2.20f	C	Mammals became the dominant large animals on Earth following an asteroid strike 66 million years ago that caused a mass extinction.
2.1i	D	*Megatherium americanum* was a very large sloth that lived in the Pleistocene.
2.2i	C	Miacids are thought to have evolved into modern day carnivores.
2.3i	C	This is a Persian onager (*Equus hemionus onager*).
2.4i	A	*Paraceratherium* was a prehistoric rhinoceros.
2.5i	A	Cetaceans are thought to have evolved from even-toed ungulates.
2.6i	D	*Australopithecus* means southern ape. The first specimen was found at Taung in South Africa.
2.7i	D	The foramen magnum is the aperture at the base of the skull through which the spinal cord passes. When positioned underneath the skull this indicates that the animal had an upright posture and was bipedal.
2.8i	B	B is a blue wildebeest (*Connochaetes taurinus*) and is not a pachyderm. The term 'Pachydermata' refers to an obsolete order of mammals. The term included thick-skinned species: elephants, hippopotamuses and rhinoceroses. Zoos used to house these species together in a Pachyderm House.
2.9i	C	A tautonym is a scientific (binomial) name in which the name of the genus and that of the species are identical. Note that in 'B' the two names sound the same but are spelt differently.
2.10i	D	The family Ornithorhynchidae contains a single species: the duck-billed platypus (*Ornithorhynchus anatinus*).
2.11i	B	The Chiroptera contains the bats and is the second most speciose mammalian order.
2.12i	D	These names indicate subspecies of the chimpanzee *Pan troglodytes*.

2.13i	A	The hyoid is a small U-shaped bone involved in breathing, speech and swallowing in humans.
2.14i	D	Rabbits and hares were previously classified as rodents.
2.15i	C	The puma has a very large range extending from the northwest United States to the southern tip of South America. This has resulted in the use of a large number of local names for this species.
2.16i	B	The Edentata is an obsolete order of mammals that contained the anteaters, sloths and armadillos. These taxa are now grouped together in the Xenarthra.
2.17i	B	The blue whale is a baleen whale.
2.18i	A	Fig. 2.6 is a model of a *Brontotherium*, an extinct rhinoceros-like browser that was endemic to North America.
2.19i	A	These are all proboscideans related to modern elephants.
2.20i	C	Chimpanzees and bonobos are the apes that are most closely related to humans.
2.1a	C	C is correct.
2.2a	D	The Laetoli footprints were discovered by Mary Leakey in northern Tanzania and were made by *Australopithecus afarensis*. These footprints were the oldest known evidence for bipedalism in hominins at the time.
2.3a	A	*Zinjanthropus* was discovered by Mary Leakey in northern Tanzania in Olduvai Gorge. This form is now known as *Paranthropus boisei* but some authorities consider this genus to be synonymous with *Australopithecus* so they call it *Australopithecus boisei*.
2.4a	D	'La Grande Coupure' means 'the great cut' and refers to a wave of extinction of mammals that occurred about 32 million years ago.
2.5a	B	Cope's rule refers to the tendency of animals in evolutionary lineages to increase in size with time. Increased size benefits survival, fecundity and mating success.
2.6a	A	*Mystacodon* is a fossil cetacean.
2.7a	D	D is correct. This suborder contains dog-like forms, including bears and raccoons.
2.8a	B	The order Microbiotheria contains a single species: the monito del monte.
2.9a	C	Fossils of woolly mammoths have been found in North America and Siberia.
2.10a	B	Growth lines in fossil teeth allow palaeontologist to deduce these aspects of their reproductive biology.
2.11a	B	B is correct.
2.12a	D	Members of the Strepsirrhini possess a tooth comb.

2.13a	A	Farnsworth *et al.* (2023) predicted a merging of the continents and a mass extinction of mammals caused by global warming.
2.14a	C	C is correct.
2.15a	A	This is a model of an entelodont: an extinct, pig-like artiodactyl.
2.16a	D	D is correct.
2.17a	B	The Vespertilionidae is the largest family of bats. It contains microbats.
2.18a	A	A is correct.
2.19a	C	The Afrotheria is a superorder containing mammals of African origin.
2.20a	C	The Diprodontia contains the kangaroos, wallabies, koala, wombats, possums and their relatives. The Tasmanian devil (*Sarcophilus harrisii*) is a member of the order Dasyuromorphia.

3. Anatomy

3.1f	C	The structure labelled 'X' in Fig. 3.1 is a diastema: a space between the teeth. The image shows the skull and upper jaw of a horse.
3.2f	A	The dentary has a single bone in mammals.
3.3f	C	Most mammals, including giraffes, possess seven cervical vertebrae. Sloths and manatees are exceptions.
3.4f	B	Lordosis is the inward curve of the lumbar spine. The term is also used for the mating posture of some sexually receptive mammals (e.g. rats) whereby the head and rump are raised and the back arched downward.
3.5f	D	An omasum is only present in ruminants.
3.6f	C	The baleen is made of the protein keratin.
3.7f	A	Pangolins (scaly anteaters) are covered in keratinous scales.
3.8f	B	B is correct.
3.9f	C	The cheetah is unusual in being unable to fully retract its claws.
3.10f	A	Carnassial teeth are slicing teeth used by carnivorans to cut meat.
3.11f	D	The structure labelled 'X' is the superior vena cava which carries venous blood from the upper (anterior) half of the body to the right atrium.
3.12f	C	The structure labelled 'Y' is the left ventricle.
3.13f	A	A is correct.
3.14f	A	The colobus does not have a prehensile tail. This is an Old World species. Prehensile tails in primates are only found in New World monkeys.
3.15f	B	Purkinji fibres are made from specialised electrically-excitable cells that conduct cardiac action potentials in the inner ventricular walls and create synchronised contractions of the ventricles.

3.16f	D	A nictitating membrane is a 'third eyelid'; a thin membrane that may be drawn across the eye for protection. Occurs, for example, in seals, polar bears and cats.
3.17f	D	D is correct.
3.18f	B	A patagium is a membrane that extends between the forelimbs and hindlimbs providing flying squirrels with a wing-like structure that allows them to glide.
3.19f	C	Turbinal bones occur in the nasal chamber of the skull.
3.20f	A	The primary function of the masseter muscle is to elevate the jaw bone, bringing the teeth together for chewing.
3.1i	C	This image shows the feet of a Bactrian camel (*Camelus bactrianus*). This animal has a digitigrade stance, i.e. it stands and walks on its toes.
3.2i	D	The image shows the toes of a two-toed sloth (*Choloepus didactylus*); a member of the order Pilosa.
3.3i	A	Muscles contain protein fibres made of actin and myosin.
3.4i	B	Only giraffes and okapis possess ossicones: a pair of protuberances on the head made of ossified cartilage and covered in skin and hair.
3.5i	C	Fig. 3.8 is an image of an intervertebral disc of a whale.
3.6i	C	C is correct.
3.7i	B	The corpus callosum connects the two cerebral hemispheres.
3.8i	A	In a typical mammal, the elbows face backward and the knees face forwards.
3.9i	A	A is correct
3.10i	D	Pikas are lagomorphs and as such possess a diastema between the incisors and check teeth. A = jaguar; B = polar bear; C = human.
3.11i	B	The aperture labelled 'X' is the foramen magnum. The skull is that of a gorilla.
3.12i	D	The tooth labelled 'X' is a carnassial tooth. The skull is that of a red fox (*Vulpes vulpes*)
3.13i	D	D is correct. Vibrissae are also known as whiskers.
3.14i	B	B is the sagittal crest. This is used for the attachment of the jaw muscles.
3.15i	A	This is a baleen from a minke whale.
3.16i	B	The clitoris in the female binturong (*Arctictis binturong*) (Fig. 3.13(i)) and the female spotted hyena (*Crocuta crocuta*) (Fig. 3.13 (ii)) is enlarged and often termed a pseudopenis. Fig. 3.13 (iii) is a Guinea baboon (*Papio papio*) and Fig. 3.13 (iv) is a maned wolf (*Chrysocyon brachyurus*).

3.17i	A	A is correct.
3.18i	D	D is correct.
3.19i	B	B is the correct sequence from head to tail.
3.20i	C	An ischial callosity is a hard 'sitting pad' found characteristically on the buttocks of Old World monkeys.
3.1a	B	B is correct. 1 = giant panda; 2 = grey wolf; 3 = gorilla; 4 = lion.
3.2a	A	Giraffes have a hard pad of tissue in the upper jaw in place of incisors. The incisors of the lower jaw bite against this pad.
3.3a	B	The dactylopatagium is the part of a bat's wing membrane that spans digits 2 to 5.
3.4a	D	D is correct.
3.5a	C	C is correct. Syndactyly (webbed feet) occurs normally in some mammals, especially the siamang and members of the Diprotodontia.
3.6a	B	These terms describe the different types of uterus found in various mammal species.
3.7a	A	The baculum is the penis bone or os penis. It is present in most primates but not humans.
3.8a	A	Dermatoglyphics are fingerprints.
3.9a	B	B is the zygomatic arch.
3.10a	D	D is correct.
3.11a	D	Sound is made by the larynx in baleen whales (mysticetes) and a nasal organ in toothed whales (odontocetes).
3.12a	A	These are regions of the stomach.
3.13a	C	This study was published by the French zoologist Georges Cuvier in 1812.
3.14a	B	The aye-aye has exceptionally long and thin middle fingers for collecting grubs and larvae from cracks and holes in tree bark.
3.15a	D	A cloaca is an organ used for expelling nitrogenous waste to the outside and for mating. A cloaca is present in all of these species.
3.16a	B	Elephant teeth are an example of lophodont dentition: cheek teeth with transverse cusps.
3.17a	A	These are processes (projections) found in some vertebrae, which articulate with similar processes on adjacent vertebrae.
3.18a	C	The Islets of Langerhans produce insulin in the liver.
3.19a	D	These are structures in the skin that detect skin stretching, finger movement and finger position (Ruffini corpuscles) and fine textures and vibrations (Pacinian corpuscles).
3.20a	B	A perineal tumescence is a sexual swelling used to attract males.

4. Physiology

4.1f	C	Saltatory means leaping or proceeding by abrupt movements.
4.2f	B	Sight is the most important primate sense.
4.3f	D	Resting heart rate is negatively correlated with size in mammals: small mammals have high heart rates.
4.4f	D	Mature mammalian red blood cells are biconcave in shape, possess no nucleus and possess the oxygen-carrying pigment haemoglobin.
4.5f	C	The Malpighian body is made up of the glomerulus and the Bowman's capsule in the kidney.
4.6f	A	The epiglottis is a flap of tissue that closes off the trachea during swallowing to prevent food and water entering the lungs.
4.7f	C	Gestation is the period between conception and birth (parturition).
4.8f	B	B is correct.
4.9f	C	The diving reflex manages the oxygen in the body while underwater: non-essential processes such as digestion shut down, blood flow is maintained to the brain and muscles, heart rate falls, breathing stops, peripheral blood vessels constrict, etc.
4.10f	A	Cells in the hypothalamus in the brain monitor blood temperature so that it may be maintained within relatively narrow limits.
4.11f	D	Arrector pili cause the hairs to stand away from the skin as a means of insulation against heat loss.
4.12f	A	This is behavioural thermoregulation because the hyrax exposes itself to the sun to warm up and when the sun is not shining reduces heat loss by huddling with others. Both of these actions involve changes in behaviour.
4.13f	C	This wombat produces cuboidal faeces.
4.14f	A	The epididymis stores sperm in the testes.
4.15f	B	The milk of marine mammals and reindeer is very rich in fat.
4.16f	A	Sebaceous glands are microscopic exocrine glands that open into hair follicles and produce an oily substance called sebum.
4.17f	A	The basal metabolic rate or resting metabolic rate is the rate of energy use required to sustain basic life functions when the body is at rest.
4.18f	B	Koalas have a low metabolic rate compared with other species of a similar size.
4.19f	C	Normal blood pressure at the heart of a giraffe is 220/180 mmHg. The high systolic pressure is required to raise the blood to the head.
4.20f	D	The term 'hygroscopic' refers to an ability to absorb moisture from the air.

4.1i	C	The kangaroo rat is a desert species. The long Loop of Henlé allows its kidneys to reabsorb more water from its urine than would otherwise be possible.
4.2i	D	Lactose is a disaccharide consisting of one molecule of glucose and one molecule of galactose.
4.3i	C	Diastole is the phase of the cardiac cycle when the heart muscles relax so that the heart chambers may fill with blood prior to the next contraction (systole).
4.4i	B	B is correct.
4.5i	B	B is correct.
4.6i	D	The Jacobson's organ (also known as the vomeronasal organ) is a paired olfactory sense organ located in the soft tissue of the nasal septum.
4.7i	A	A tendon and ligament spring stores energy during running.
4.8i	A	This squirrel allows its body temperature to fall below 0°C during torpor.
4.9i	D	Luteinising hormone level peaks in the blood immediately prior to ovulation.
4.10i	B	Meissner's corpuscles are concerned with touch and occur in the hands and feet of primates.
4.11i	C	These are all types of pheromones.
4.12i	B	B is correct.
4.13i	A	The pygmy slow loris produces toxins from modified sweat glands near its elbows. It licks these glands when alarmed and transfers the toxin to its teeth.
4.14i	B	The corpus luteum is a temporary organ that forms in the ovary and disappears if fertilisation does not occur. In early pregnancy it produces the hormone progesterone to maintain the embryo.
4.15i	A	Altricial means born helpless and needing considerable parental care.
4.16i	D	D is correct.
4.17i	C	Blood vessel 'X' is the hepatic portal vein. It carries blood from the gut to the liver.
4.18i	D	D is correct.
4.19i	A	The hypothalamic-pituitary-adrenal axis (HPA) is the body's stress response system.
4.20i	D	D is correct. Golden moles reduce their need for food and water by reducing their physical activity and maintaining a low body temperature and low metabolic rate.
4.1a	A	Dehnel was a Polish zoologist.
4.2a	B	B is correct.

4.3a	C	In marine mammals red cell numbers in blood are twice the level in terrestrial mammals and myoglobin is nine times as high.
4.4a	A	Semelparity is the capacity to breed just once in an animal's lifetime.
4.5a	C	Eimer's organ is a touch receptor in the snouts of moles and desmans.
4.6a	B	These structures make a sound when rubbed together.
4.7a	A	The forelimb-hindlimb phase relationship describes the pattern of movement of the legs in the gait of quadrupeds.
4.8a	B	Larger mammals take longer to enter torpor than do smaller mammals.
4.9a	D	The spermaceti organ is located in the head and consists of liquid waxes.
4.10a	C	This is the Bohr effect or shift.
4.11a	D	D is the correct sequence.
4.12a	D	D is correct.
4.13a	C	Embryonic diapause allows macropods to delay the development of the embryo.
4.14a	A	A represents the oxygen dissociation curve of myoglobin. This pigment is found in cardiac and skeletal muscles and becomes fully saturated with oxygen at a lower partial pressure of oxygen than haemoglobin.
4.15a	A	A deciduate mammal sheds an afterbirth at parturition.
4.16a	C	The platypus releases toxins via spurs on its rear legs.
4.17a	C	C is correct.
4.18a	A	Solenodons are shrew-like insectivorous mammals.
4.19a	B	The red pigment contains an antibiotic, the orange pigment absorbs ultra violet light and acts as a sunscreen.
4.20a	D	This structure is particularly important in nocturnal species.

5. Behaviour

5.1f	B	A facial threat is an aggressive behaviour. Affiliative behaviour consists of friendly and peaceful acts.
5.2f	C	A large social group of spotted hyenas (*Crocuta crocuta*) is called a clan.
5.3f	A	The term 'fossorial' refers to the habit of burrowing underground.
5.4f	D	The image shows a black-tailed prairie dog (*Cynomys ludovicianus*) standing next to the entrance to a burrow. This species lives in colonies (or 'towns') that occupy a system of interconnected underground burrows with multiple entrances.
5.5f	A	In most mammal species it is the juvenile males that disperse from the natal group.

5.6f	D	An ethogram is a detailed description of the behavioural repertoire of a species.
5.7f	D	Appeasement behaviour in mammals may take a variety of forms including all of the behaviours described in A, B and C.
5.8f	C	C is correct.
5.9f	B	This is stereotypic behaviour because it is repetitive, invariant in form and has no apparent purpose. This type of behaviour is also called route-tracing behaviour when the animals repeatedly follow the same route. Many species living in zoos exhibit this type of route-tracing behaviour, especially elephants, big cats, bears and other predators.
5.10f	C	This occurs in vampire bats. A well-fed bat may regurgitate blood to feed a relative or 'friend'.
5.11f	B	B is correct.
5.12f	B	Grooming others is an important behaviour that helps to establish and maintain social relationships, especially in primates.
5.13f	A	The naked mole-rat (*Heterocephalus glaber*) is a burrowing rodent. It may be described as eusocial as it consists of reproductive and non-reproductive groups (castes), and exhibits cooperative brood care.
5.14f	C	African wild dogs (*Lycaon pictus*) perform an elaborate greeting ceremony prior to hunting. The other species are: A = Iberian wolf (*Canis lupus signatus*); B = spotted hyena (*Crocuta crocuta*); D = African lion (*Panthera leo*).
5.15f	D	All of these species adopt defensive formations when threatened.
5.16f	D	Elephants, sea otters (Mustelidae) and oceanic dolphins (Delphinidae) have been observed using tools.
5.17f	C	This is a type of social behaviour whereby individuals are stimulated to engage in a behaviour being performed by others located nearby.
5.18f	A	Springbok adopt a 'bouncing' gait when alarmed by extending all four legs simultaneously.
5.19f	B	A rapid series of clicks produced by a dolphin is called a 'click train'.
5.20f	D	This heavily used area is the 'core area'.
5.1i	A	Harlow used Rhesus monkeys.
5.2i	C	The term 'philopatry' means staying in the area where it was born.
5.3i	A	Mt Elgon is on the border of Kenya and Uganda.
5.4i	C	Ring-tailed lemurs (*Lemur catta*) engage in stink fights. A male who feels threatened will spread secretions from glands on his hands onto his tail and wave it in the air to spread the odour. The other species illustrated are: A - De Brazza's monkey (*Cercopithecus neglectus*); B - Eastern pygmy marmoset (*Cebuella niveiventris*); D - black and gold howler monkey (*Alouatta caraya*).

5.5i	B	B is correct.
5.6i	D	Humpback whales catch fish by creating a 'net' of bubbles.
5.7i	B	A cathemeral animal is one that is active throughout various times of the day.
5.8i	A	Releaser pheromones trigger behaviours in other individuals. For example, female rabbits release mammary pheromones that trigger nursing behaviour in their young.
5.9i	C	C is correct.
5.10i	A	A is correct.
5.11i	B	A circadian rhythm is based on a 24-hour cycle. A circannual rhythm is based on an annual cycle.
5.12i	D	Maned wolves do not hunt in packs and take only small prey as part of an omnivorous diet.
5.13i	D	*Kanzi* and *Panbanisha* are bonobos who participated in a number of studies of language at Georgia State University.
5.14i	B	The term 'scansorial' means adapted for, or capable of, climbing.
5.15i	C	This behaviour is part of the mating behaviour of hooded seals.
5.16i	D	In a fission-fusion society individuals come together in social groups and move between groups.
5.17i	C	Alloparenting is a behaviour whereby parental care is given by an adult to a young individual who is not their offspring.
5.18i	D	D is correct.
5.19i	B	Katy Payne and her co-workers first reported the use of infrasound by Asian elephants at Washington Park Zoo.
5.20i	D	Howler monkeys use all of these types of calls. Fig. 5.6 is a group of black and gold howler monkeys (*Alouatta caraya*).
5.1a	D	This is the Bruce effect.
5.2a	C	The leg beat is a courtship behaviour in roan antelopes.
5.3a	A	This behaviour was reported in an orangutan in Laumer *et al.* (2024).
5.4a	D	Packer and Pusey (1982) established that around 50% of coalitions of male African lions included at least one individual who was not related to the others. This is important because we would expect cooperating individuals to be related.
5.5a	A	The Prisoner's Dilemma uses a mathematical model to assess the relative pay-offs when two individuals co-operate in a task compared with when they do not.
5.6a	B	Rutting behaviour in red deer is triggered by a change in day length.
5.7a	B	*Happy* was an Asian elephant.

5.8a	C	Many wild mammals have altered various aspects of their behaviour and ecology as an adaptation to living in urban areas.
5.9a	D	Homosexual behaviour is particularly common in primates.
5.10a	A	Caribou, or reindeer, (*Rangifer tarandus*) remain in auditory contact via the sounds made by their foot bones as they walk. The other species are B: red deer (*Cervus elaphus*); C – Hartmann's mountain zebra (*Equus zebra hartmannae*); D – Dorcas gazelle (*Gazella dorcas*).
5.11a	C	Verraux's sifakas (*Propithecus verreauxi*) cool down by hugging trees. They hug the bottom of the trunk as this may be 5°C cooler than the surrounding air on hot days (Chen-Kraus *et al.*, 2023).
5.12a	C	C is correct.
5.13a	B	This is sequential assessment.
5.14a	A	This is a change in syntax.
5.15a	C	This is reciprocal altruism. This is altruism that takes place between unrelated individuals when there will be (or is likely to be) repayment of the altruistic act at some time in the future.
5.16a	B	The experience of the signaller is the most important factor in determining the extent to which individuals use the social information they provide.
5.17a	D	This is a mechanism by which individuals may recognise relatives they have not previously encountered. This depends upon the existence of a correlation between phenotypic similarity and genetic similarity.
5.18a	A	The blackbuck is an antelope native to India and Nepal. Males set up territories called leks to attract females for mating.
5.19a	B	A sociogram is a type of network diagram showing the degree of association between members of a group. For example, in a group of three individuals A, B and C a sociogram would take the form of a triangle made up of lines joining A to B, B to C and C to A. The thickness of each line would be proportional to the amount of time each dyad (pair) spends associating together.
5.20a	D	Male hyraxes use a rhythmic song to advertise their quality to prospective mates.

6. Ecology and Genetics

6.1f	C	C is correct.
6.2f	B	Such a species eats snakes, e.g. a mongoose.
6.3f	A	Fig. 6.1 (A) shows two Asian short-clawed otters (*Aonyx cinereus*). Riparian means relating to wetland adjacent to rivers and streams. B = red panda (*Ailurus fulgens*); C = Hamadryas baboon (*Papio hamadryas*); D = orangutan (*Pongo* sp.)

6.4f	B	Male bonobos have an 'X' and a 'Y' chromosome, the same as human males and most male mammals.
6.5f	D	According to the Food and Agriculture Organisation (2020) 68% of all mammal species live in forests.
6.6f	C	A black male would have the genotype ww. A white male whose mother was black would have the genotype Ww. The result of this mating would produce the following offspring:
6.7f	B	Beavers are considered ecosystem engineers because of their dramatic effects on the ecology of the habitats in which they live.
6.8f	D	These ecological changes were the result of the reintroduction of grey wolves after a long period of absence.
6.9f	C	The male is the heterogametic sex possessing one of each of the two types of sex chromosomes, X and Y.
6.10f	A	Figs 6.2(ii) and (iii) show generalists; a Canada lynx (*Lynx canadensis*) and a raccoon (*Procyon lotor*). Fig. 6.2(i) is a koala (*Phascolarctos cinereus*) and Fig. 6.2(iv) is a giant anteater (*Myrmecophaga tridactyla*).
6.11f	B	The aardvark is a secondary comsumer or a tertiary producer and is myrmecophagous. An osteophage eates bones. Decomposers are bacteria and fungi involved in decomposition.
6.12f	B	All of these species are small antelopes. The gerenuk has a long neck and can reach high in the vegetation when it stands on its hind legs.
6.13f	D	The majority of the mammalian biomass on Earth is made up of livestock and humans.
6.14f	B	These extra chromosomes are called B chromosomes.
6.15f	C	Life expectancy is a prediction. The elephant could live up to another 55 years (70-15).
6.16f	C	The British ecologist Charles Elton established the Bureau of Animal Population at the University of Oxford in 1932 which became an important centre for the study of fluctuations in animal populations.
6.17f	B	This is fecundity rate.
6.18f	A	A leucistic lion would have blue eyes and white fur. This condition is inherited and is not the same as albinism.
6.19f	D	A cat with Klinefelter's syndrome is a male with an extra X chromosome (XXY).

For answer 6.6f, the Punnett square is:

		Male gametes	
		w	w
Female gametes	W	Ww	Ww
	w	ww	ww

This means that 50% of the offspring would be black (ww).

6.20f	A	A is correct.
6.1i	D	D is correct. Statement 'v' is not true as desert rodents avoid being active during the hottest period of the day.
6.2i	C	*K*-selected species are large, long-lived and have few offspring.
6.3i	D	This is the Allee effect.
6.4i	A	A Skinner trap is used for catching moths. Sherman and Longworth traps are used to catch small terrestrial mammals. A harp trap is used to capture bats.
6.5i	D	D is correct. The other indices are fictitious.
6.6i	A	This relationship is an example of mutualism in which both organisms benefit.
6.7i	C	Some pikas feed on flowers.
6.8i	B	These populations are allopatric, that is they do not overlap. Asiatic lions occur in northwest India and African lions live in savannah habitats south of the Sahara.
6.9i	D	Dolly the sheep was the first mammal to be cloned from an adult somatic cell. Her body was stuffed and preserved and is held by the National Museum of Scotland in Edinburgh.
6.10i	A	Crepuscular animals are active at twilight.
6.11i	C	African elephants have the lowest intrinsic rate of natural increase (*r*) of the species listed. Gestation is approximately 22 months and elephants usually only produce a single offspring although twins occur very rarely.
6.12i	A	In polygynous mating systems a single male mates with many females.
6.13i	C	A horro is a breed of cattle. A tigon is the result of a cross between a male tiger and a female lion; a liger is the result of a cross between a male lion and a female tiger; a zonkey is a cross between a zebra and a donkey.
6.14i	D	A polygene is any of a group of genes that together control the inheritance of a quantitative character.
6.15i	A	A is correct. Mortality in humans is low until old age and then it increases dramatically.
6.16i	C	Members of the Megachiroptera feed on fruit.
6.17i	A	A is correct.
6.18i	B	B is correct.
6.19i	D	Caughley examined the relationship between cycles in elephant numbers and changes in the abundance of the trees on which they feed.
6.20i	D	Sirenians also occur in freshwater. The Amazonian manatee (*Trichechus inunguis*) lives in the Amazon basin.

6.1a	C	The platypus has five pairs of sex chromosomes.
6.2a	B	The aardvark disperses the seeds of the cucurbit plant.
6.3a	D	The horns display annual rings that indicate the age of the sheep.
6.4a	B	Any deviation from a 50:50 ratio reduces the effective population size.
6.5a	A	The Y-chromosome is smaller than the X-chromosome.
6.6a	C	C is correct.
6.7a	A	This effect was first described by the American ecologist Charles J. Krebs.
6.8a	B	Marion Island is in the southern Indian Ocean and is part of South Africa.
6.9a	A	This is a Leslie matrix named after Patrick H. Leslie.
6.10a	C	C is correct
6.11a	D	In this species both sexes possess a single X chromosome and are therefore XO.
6.12a	A	The presence of large mammals in temperate, subtropical and tropical grassland ecosystems causes a *decrease* in bush and forest fires.
6.13a	C	Mammalian megafauna are defined as those species greater than 45 to 1000 kg for herbivores and 15 to 100 kg for carnivores.
6.14a	A	This is Chitty's hypothesis.
6.15a	B	As the density of reindeer increases calf production falls.
6.16a	D	D is correct.
6.17a	D	The index assumes that the size of the population does not change between the time when the first sample is taken (and marked) and the time when the second sample is taken.
6.18a	B	Baleen whales have the largest global biomass.
6.19a	A	A is correct.
6.20a	C	C is correct. As one species increase the other decreases. This *may* be evidence of interspecific competition but the relationship may be due to other factors.

7. Zoogeography

7.1f	A	Wallace's line separates the Oriental and Australasian faunal regions. This is a theoretical boundary that separates the distinctive faunas of Asia and Australasia.
7.2f	C	There are no bears in Africa.
7.3f	D	The dhole (*Cuon alpinus*) is a highly social canid native to Central, East, South and Southeast Asia.
7.4f	B	Myers was a British environmentalist who specialised in biodiversity and was an advisor on environmental issues to the United Nations, World Bank and many other institutions.

7.5f	A	Capuchins are New World monkeys.
7.6f	B	This is a filter route as it only allows certain types of species to pass through, in this case those adapted to cold environments.
7.7f	A	Of the four areas indicated on the map, excluding introduced species, the natural distributions of marsupials and terrestrial placentals only overlap in South and Central America (area A).
7.8f	A	Africa is home to large numbers of bovids from over 70 species, almost all of which are antelopes.
7.9f	D	Mammals are able to occupy a wide range of habitats because they are able to maintain a more-or-less constant internal body temperature.
7.10f	A	Indonesia has the largest number of species of mammals.
7.11f	B	The Tibetan fox (*Vulpes ferrilata*) survives at high altitude in the Tibetan Plateau at heights of up to 5,200m.
7.12f	C	Barbary macaques are native to the Atlas Mountains of North Africa.
7.13f	D	A cline is a measurable gradient in a particular character of a species across its geographical range. In species that occupy a large range this results because individuals have a limited capacity to disperse.
7.14f	B	Naked mole-rats live in eastern Africa.
7.15f	C	C is correct.
7.16f	B	The fossa (*Cryptoprocta ferox*) is endemic to Madagascar.
7.17f	A	Allen's rule states that animals adapted to warm climates have longer and thinner limbs and body appendages (including ears) than those adapted to cold climates. Fennec foxes have unusually large ears.
7.18f	D	Africa has not experienced a mass extinction of its megafauna as evidenced by the wide range of large mammals living in Africa today.
7.19f	B	Of the regions listed, North America has the lowest proportion of endemic mammal species.
7.20f	A	Apart from bats and rodents, Australia has no native terrestrial eutherian species.
7.1i	C	These animals are pudu (*Pudu puda*) from South America.
7.2i	A	Moose live in taiga or boreal forest.
7.3i	B	Maned wolves live in the Pantanal – a large flooded grassland – in South America.
7.4i	D	The Canada lynx (*Lynx canadensis*) is native to northern North America and therefore a Nearctic species.
7.5i	D	Only wildcats (*Felis silvestris*) are native to the British Isles.
7.6i	A	A is correct.
7.7i	A	Ngorongoro crater in Tanzania has a very high density of ungulates, especially wildebeest, plains zebra, Thomson's gazelle, Grant's gazelle, and buffalo and large numbers of predators, especially lions and spotted hyenas.

7.8i	B	Numbats are insectivorous marsupials and their range is restricted to two small areas in Western Australia.
7.9i	C	The Xenarthra consists of sloths, armadillos and anteaters. They are found in an area that extends from the central United States to the southern tip of South America.
7.10i	D	The spectacled or Andean bear (*Tremarctos ornatus*) occurs in the Andes in northern and western South America. The neotropical region extends south from Mexico.
7.11i	A	Golden moles occur in sub-Saharan Africa.
7.12i	C	The sirenians consist of the dugongs and the manatees. None of the extant species is found in the Mediterranean Sea.
7.13i	D	The greatest biodiversity within the primates is seen in the neotropical region and in Asia.
7.14i	C	Pottos are small primates that inhabit the rainforests of West and Central Africa.
7.15i	A	A is correct.
7.16i	D	Bovids (antelopes, buffalo, bison, cattle, sheep, goats and their relatives)
7.17i	A	Continental drift is the moving of continents relative to each other over geological time.
7.18i	C	C is correct.
7.19i	A	Kinkajous are small rainforest carnivores native to Central and South America.
7.20i	D	The taxa in option D occur in South America.
7.1a	D	D is correct.
7.2a	A	The Comoro Islands are located in the Indian Ocean between Mozambique and Madagascar.
7.3a	C	The natural distribution of the platypus is confined to the east coast of mainland Australia and Tasmania.
7.4a	C	South-east Arizona has the highest mammal species richness.
7.5a	B	B is correct.
7.6a	D	Tarsiers are small primates that are only found on various islands in Southeast Asia.
7.7a	D	The greater cane rat is native to much of Africa south of the Sahara.
7.8a	A	The neotropics support a large number of bat and rodent species.
7.9a	C	C is correct.
7.10a	B	Alfred Russel Wallace was the first to notice this (Wallace, 1852).
7.11a	A	Gloger's rule is named after Constantin Gloger, a German zoologist. He published this theory in 1833.

7.12a	D	There are no endemic families of mammals in the Palaearctic region.
7.13a	A	Philip J. Darlington Jr was an American zoologist who theorised that the dominant vertebrate groups originated in the tropics of the Old World and explored this in his book *Zoogeography: The Geographical Distribution of Animals*, in 1957.
7.14a	C	The Isthmus of Panama arose from the sea floor as a result of volcanic activity, joining the two land masses, and thereby allowing the movement of animals in both directions.
7.15a	B	In almost all families bat species diversity increases north to south so the best predictor is latitude, with fewer species in high latitudes and more species in the tropics.
7.16a	D	The niches are larger – not smaller – on islands due to reduced interspecific competition.
7.17a	D	Volant is an adjective that means engaged in, or having the power of, flight.
7.18a	B	Vicariance is the name of the process by which the range of a species becomes split while remaining in place as a result of tectonic movements, geological activity or as a result of some other process. The emergence of the Isthmus of Panama – creating a land bridge between North and South America – separated the marine organisms on either side of Central America.
7.19a	A	Species belonging to these families are endemic to the Nearctic.
7.20a	C	A nunatak is an isolated area of rock that projects above ice or snow and may act as a refuge for species during a glaciation.

8. Conservation and Management

8.1f	B	A species that has been classified as 'least concern' by the IUCN has been evaluated and determined not to require conservation measures because it is abundant in the wild.
8.2f	A	The quagga is an extinct form of equid the rear of whose body was striped in the manner of a zebra.
8.3f	B	In 2022 the IUCN recognised 6,577 extant species of mammals.
8.4f	D	The mountain gorilla has been well protected in the last 50 years in the wild by armed guards and the health of the population has been improved by veterinary interventions.
8.5f	B	The IUNC list is the Red List.
8.6f	C	The WWF logo is a giant panda. The original concept for the logo was designed by the conservationist Sir Peter Scott.
8.7f	B	The greater one-horned rhinoceros occurs in Kaziranga National Park in Assam, India.

8.8f	A	Panthera works to conserve species of wild cats and their habitats.
8.9f	B	The black-footed ferret (*Mustela nigripes*) has been saved from extinction by a captive breeding programme operated by the US Fish and Wildlife Service and its partners. The species was thought to be extinct but was rediscovered in Wyoming in 1981.
8.10f	A	A is correct according to data published by the International Rhino Foundation (2023).
8.11f	B	The availability of food regulates most very large mammal populations because they have no natural predators.
8.12f	B	Inbreeding depression occurs when close relatives mate causing deleterious recessive genes to occur more frequently in homozygotes so that their effect manifests itself in the phenotype.
8.13f	C	C is the correct sequence. Below 'Near Threatened' a species may be described as 'Least Concern', 'Data Deficient' and 'Not Evaluated'.
8.14f	D	The Action Plan identified all six of these threats.
8.15f	A	The Sundarbans National Park is in West Bengal in India. The mangrove forest here is home to the Ganges River dolphin and the Irrawaddy dolphin.
8.16f	B	B is correct.
8.17f	C	The red lechwe is an endangered wetland antelope species of south-central Africa. Of the parks listed it only occurs in Chobe in Botswana. Korup is in Cameroon. Kaziranga and Corbett are in India.
8.18f	B	The elephant population of Tsavo suffered devastating losses of elephants in the early 1970s due to drought.
8.19f	A	The Bison Range is located in Montana.
8.20f	D	Myxomatosis is caused by a poxvirus that causes high mortality in European rabbits.
8.1i	B	A population of European bison survives in the Białowieża Forest on the border between Poland and Belarus.
8.2i	C	A moratorium on commercial whaling was established in 1986.
8.3i	A	All National Parks in South Africa are fenced.
8.4i	B	Operation Oryx was a project to prevent the extinction of the Arabian oryx.
8.5i	B	This is the Agreement on the Conservation of Polar Bears 1973.
8.6i	C	European beavers have been reintroduced into Scotland under licence.
8.7i	D	This is one of several panda breeding centres in China.
8.8i	A	The vaquita is a species of porpoise found only in the northern Gulf of California.
8.9i	D	The last quagga died in Artis Zoo in Amsterdam in 1883.
8.10i	A	This echidna was recorded by camera traps in Indonesia.

8.11i	C	C is correct.
8.12i	D	This phenomenon is known as an extinction debt. This process may take hundreds of years.
8.13i	B	The surrogate species was an eland (*Taurotragus oryx*). A = impala (*Aepyceros melampus*); C = scimitar-horned oryx (*Oryx dammah*); D = greater kudu (*Tragelaphus strepsiceros*).
8.14i	C	The Caspian, Javan and Balinese subspecies of tiger were considered to be extinct by the IUCN/SSC Red List of Threatened Species at the end of 2023.
8.15i	C	C is correct.
8.16i	D	D is correct.
8.17i	D	The first gorilla born in captivity was *Colo*, a Western lowland gorilla born in Columbus Zoo, Ohio, United States in 1956.
8.18i	C	Most attempts to use parasites to control mammal populations have failed, with the notable exception of the use of myxomatosis to control rabbits in Australia.
8.19i	B	USDA kills very large numbers of coyotes.
8.20i	A	A stochastic event is a random event which, by definition, in not predictable.
8.1a	A	This process has been termed 'Pleistocene rewildling' as this was the geological period when mammalian megafauna dominated the Earth.
8.2a	A	This reintroduction was controversial because of the opposition from ranchers concerned about wolves taking their livestock. The descendants of these wolves are now a major tourist attraction.
8.3a	B	The huemul (*Hippocamelus bisulcus*) is an endangered species of deer.
8.4a	A	The park is also known as the Addo Elephant National Park.
8.5a	C	The first studbook created for a wild animal species was for the European bison (*Bison bonasis*), established in 1932.
8.6a	D	The minimum viable population is an estimate of the minimum number of individuals required to ensure a high probability of the survival of the population over a fixed period of time
8.7a	A	Vynne *et al.* (2022) claimed that this could be achieved with just 20 species. For example, they argued that the reintroduction or recolonisation of European bison (*Bison bonasis*), grey wolves (*Canis lupus*), Eurasian beaver (*Castor fiber*), reindeer (*Rangifer tarandus*) and lynx (*Lynx lynx*) in Europe could create historically intact mammal assemblages in a further 35 ecoregions.
8.8a	C	This species has been widely reported as the first mammal species to become extinct as a result of human-induced climate change. This came about due to the inundation of the animal's habitat by the ocean.

8.9a	D	This population crash and subsequent recovery described the population changes experienced by blue whales.
8.10a	C	C is correct.
8.11a	B	Steller's sea cow was first described by the German zoologist Georg Wilhelm Stellar in 1741. It became extinct in 1768.
8.12a	B	Outbreeding depression is a negative effect of interbreeding previously isolated populations.
8.13a	D	D is correct.
8.14a	A	The overpass is known as the Wallis Annenberg Wildlife Crossing.
8.15a	B	B is correct. PZP is an abbreviation for porcine zona pellucida.
8.16a	C	C is correct. Note that this is 26% of *assessed* species not *all* species.
8.17a	A	A is correct.
8.18a	D	All of the listed species have been cloned.
8.19a	C	These are *K*-strategists. '*K*' is the symbol used for the asymptote of a population that exhibits logistic growth: its carrying capacity.
8.20a	B	Fences cause disruptions to migrations of some species, cause population fragmentation and kill animals as a result of entanglement.

9. Parasites and diseases

9.1f	B	The adults of these tapeworms live in mammalian guts.
9.2f	C	Ulcerative pododermatitis (bumblefoot) is cause by bacteria (*Staphylococcus*, *Pseudomonas* and *Escherichia*) and affects the feet. It affects rodents, rabbits and birds.
9.3f	D	This species of *Plasmodium* cause malaria in some non-human primates.
9.4f	A	This organism is a trematode parasite that is transmitted by freshwater snails.
9.5f	B	*Taenia* is a genus of tapeworm.
9.6f	D	Any mammal can be infected with rabies.
9.7f	A	*Yersinia pestis* causes plague: the Black Death.
9.8f	C	Foot-and-mouth disease predominantly affects cloven-hoofed mammals (e.g. sheep, cows, goats, pigs, llamas, deer, buffalo and their relatives).
9.9f	D	D is correct. Sheep are ruminants and ruminants are the principal definitive host of this parasite.
9.10f	A	A coloboma is an area of missing tissue in the eye.
9.11f	C	C is correct.
9.12f	D	B_{12} is necessary for the production of red blood cells.
9.13f	B	B is correct.

9.14f	D	Bovine mastitis affects the udder of cows.
9.15f	A	Toxoplasmosis is caused by the protozoan parasite *Toxoplasma gondii*. It is transmitted by cat faeces. The cats become infected by eating infected birds, rodents and other small animals.
9.16f	A	FeLV is a viral infection of cats.
9.17f	D	All of these conditions may be caused by obesity.
9.18f	C	Myocarditis is inflammation of the heart muscle.
9.19f	B	Cortisol is released into the blood when an animal is stressed and its metabolites appear in faeces.
9.20f	C	Sickle cell anaemia confers some protection against malaria.
9.1i	A	Simian foamy virus (SFV) is a retrovirus. It affects non-human primates such as chimpanzees and macaques, and has also been transmitted to humans working closely with non-human primates in zoos, medical research institutions and animal sanctuaries.
9.2i	B	Leptospirosis (Weil's disease) is a bacterial disease.
9.3i	D	Rinderpest (cattle plague) was an important viral disease of wild and domesticated species with cloven hooves. It was declared to be officially eradicated in 2011.
9.4i	B	Spondylosis is a degenerative disease of the vertebrae and intervertebral discs which is common in elderly bears.
9.5i	D	Monkeypox (M-pox) is a viral disease that may infect a wide range of mammals in addition to primates.
9.6i	A	CWD is caused by a prion. This is a type of protein that can cause normal proteins in the body to fold abnormally causing cell death.
9.7i	D	Orf virus is a parapoxvirus that principally infects sheep and goats.
9.8i	A	Rift Valley fever is a viral disease.
9.9i	D	EEHV is elephant endotheliotropic herpesvirus.
9.10i	B	Laminitis principally affects the hooves of equids, but other species with hooves may also be affected.
9.11i	C	Sarcoptic mange (scabies) is caused by a mite: *Sarcoptes scabiei*.
9.12i	B	This condition is called lumpy jaw.
9.13i	C	C is correct.
9.14i	A	The red fox is the most important definitive host of this parasite.
9.15i	D	Aujeszky's disease (pseudorabies) is a viral disease.
9.16i	B	Captive elephants are trained to provide samples from inside their trunks in order that they may be tested for tuberculosis.
9.17i	D	Giardiasis is a protozoan disease. Phocine distemper and seal pox are caused by viruses; leptospirosis is a bacterial disease.

9.18i	A	Capture myopathy occurs in ungulates as a result of stress after they have been pursued, captured or transported.
9.19i	D	Humans may be infected by more than 150 species of macroparasites.
9.20i	C	C is correct.
9.1a	C	Leishmaniasis is transmitted by phlebotomine sand fly, e.g. *Phlebotomus papatasi*.
9.2a	A	A paratenic host is one that is an optional intermediate host which parasites enter passively along with ingested food.
9.3a	A	Tasmanian devil populations have been depleted as a result of infection with devil facial tumour disease (DFTD). This is an aggressive non-viral, transmittable parasitic cancer.
9.4a	D	Australian bat lyssavirus is related to the rabies virus and is spread to humans via infected bats.
9.5a	C	Hepatopathy is a diseased or abnormal state of the liver.
9.6a	D	D is correct.
9.7a	D	This affects equids.
9.8a	A	A is correct.
9.9a	C	All the statements are true except iv. PCBs *have* been found in killer whales, bottlenose and striped dolphins at concentrations that are high enough to cause health damage (Jepson *et al*., 2016).
9.10a	D	All of these species, and a number of others, have been infected with COVID-19.
9.11a	D	D is correct.
9.12a	B	Sacbrood is a viral disease of honey bees.
9.13a	A	Trichinosis is a roundworm infection. The other diseases are tick-borne: Babesiosis is caused by a protozoan, anaplasmosis and tularaemia are caused by bacterial infections.
9.14a	B	A pressure plate is a piece of equipment used to examine the foot.
9.15a	C	Allostatic load is the burden of chronic stress on the body.
9.16a	A	Bears living in zoos at the time of this study had chewed on metal bars and other structures resulting in dental damage.
9.17a	D	Poorly-designed hay racks allowed giraffes to trap their jaws and fracture their mandibles. Fig. 9.3 shows several types of feeding devices for giraffes but is not intended to suggest that any of these are poorly designed or responsible for mandibular fractures.
9.18a	C	Rhabdiasis is a lung disease of amphibians caused by infection by a nematode, the lungworm *Rhabdias*.
9.19a	A	A is correct.
9.20a	B	Fruit bats are thought to be the natural hosts and principal animal reservoir of the Ebola virus.

10. Domestication and the Human Use of Mammals

10.1f	D	A Dexter is a breed of cattle.
10.2f	A	Asian elephants have been used by humans for hundreds of years but have not been selectively bred and adapted for human use. Historically, most of the elephants used by humans have been taken from the wild. African elephants have also been used by humans but, again, have not been selectively bred.
10.3f	D	All have been used as pack animals.
10.4f	C	These are all dog breeds.
10.5f	D	A Bagot is a breed of goat.
10.6f	C	C is correct.
10.7f	A	This is a Shire horse.
10.8f	B	B is correct.
10.9f	B	The first mammal to orbit the Earth was a dog called *Laika* aboard the Soviet spacecraft Sputnik 2 in 1957.
10.10f	A	Castoreum is used by beavers in scent marking.
10.11f	B	Aurochs (*Bos primigenius*) are believed to have been the ancestors of all domestic cattle breeds.
10.12f	D	Armadillos are used in leprosy research.
10.13f	C	A burro is a small donkey derived from the African wild ass (*Equus africanus*). It may be considered a subspecies of this species but is sometimes classified as the separate species *E. asinus*.
10.14f	D	The Welsh Cob is not a heavy horse.
10.15f	B	A warrener raised rabbits in an underground system of burrows called warrens.
10.16f	D	The ancestors of the modern horse evolved on the Eurasian steppes.
10.17f	C	There are approximately 800 cattle breeds.
10.18f	B	B is correct.
10.19f	A	Dalmatians were used as carriage dogs. They ran alongside horse-drawn carriages to protect the occupants from being held up by highwaymen.
10.20f	D	D is correct. These animals were domesticated in the chronological sequence dog, sheep, cow, horse.
10.1i	B	B is correct.
10.2i	C	The prefix 'onco' is derived from the Greek for 'mass' or 'tumour'.
10.3i	C	British deer farms raise red deer (*Cervus elaphus*) for meat.
10.4i	D	D is correct.
10.5i	D	All of these breeds are suitable.

10.6i	B	The Abyssinian has short hair.
10.7i	A	Golden hamsters originated in Syria.
10.8i	D	D is correct.
10.9i	C	The cavy, or Guinea pig, is farmed for food in Ecuador.
10.10i	B	Some domesticated species live a free-ranging life in feral populations, such as feral cats and feral dogs.
10.11i	C	In 1784 Lazzaro Spallanzi successfully inseminated a dog.
10.12i	D	Carabeef is water buffalo meat. Squab is a type of pigeon meat; chevon is goat meat; venison is deer meat.
10.13i	B	The Sámi are the indigenous people of Norway, Sweden, Finland and the Kola Peninsula of Russia.
10.14i	A	A is correct.
10.15i	B	The Leonberger is a large breed of dog from Leonberg in Germany.
10.16i	D	These countries and others including all Member States of the European Union have banned experimentation on great apes.
10.17i	D	D is correct.
10.18i	B	The Himalayan cat is a long-haired breed similar to a Persian. Bengal cat = domestic cat (spotted Egyptian Mau) x Asian leopard cat; savannah cat = domestic cat x serval; cheetoh = Bengal x Ocicat.
10.19i	C	Over 70 percent of bull dogs develop hip dysplasia.
10.20i	A	The Droughtmaster is an Australian cattle breed produced by crossing *Bos taurus* with *B. indicus* to overcome the problems of drought and cattle ticks in North Queensland.
10.1a	B	B is correct.
10.2a	A	A is correct.
10.3a	D	Pacas are rodents.
10.4a	A	The Shiba Inu is a Japanese dog breed originally bred for hunting.
10.5a	B	B is correct. These mummified dogs were thought to act as intermediaries between the divine and human worlds.
10.6a	A	A is correct.
10.7a	B	The book was published by Darwin in 1868.
10.8a	C	Silver foxes – melanistic red foxes – were used in a long-term breeding study that began in 1959 and was conducted in Russia by the geneticist Dmitri Belyaev.
10.9a	B	The study was conducted in Siberia.
10.10a	A	The Brokpa are an Asian ethnic minority tribe.
10.11a	D	Directional selection involves the breeding of animals so as to favour a single phenotype (in this case high milk yield) causing the allele frequency to shift continuously in one direction.

10.12a	C	Lanolin is obtained from sheep.
10.13a	D	D is correct.
10.14a	B	B is correct.
10.15a	A	According to Understanding Animal Research (2022) 0.15% of experimental procedures in Great Britain were conducted on monkeys. The use of primates in research has declined in recent years due to welfare and ethical considerations.
10.16a	D	Products such as shampoo were safety tested by dropping it into the eyes of restrained rabbits.
10.17a	A	The US Navy employs bottlenose dolphins and California sea lions in its Marine Mammal Program.
10.18a	C	Wild red kangaroos are harvested in Australia and their meat is used for human consumption and for pet food.
10.19a	C	Domestication usually results in a *decrease* in an animal's body size compared with its wild ancestors not an increase.
10.20a	A	These materials are whale products.

References

Abramsky, Z. (1981) Habitat relationships and competition in two Mediterranean *Apodemus* spp. *Oikos*, 36(2), 219-225.

Anderson, E.W. and Scherzinger, R.J. (1975) Improving quality of winter forage for elk by cattle grazing. *Rangeland Ecology & Management/Journal of Range Management Archives*, 28(2), 120-125.

Caughley, G. (1976) The elephant problem–an alternative hypothesis. *African Journal of Ecology*, 14(4), 265-283.

Ceballos, G. and Ehrlich, P.R. (2006) Global mammal distributions, biodiversity hotspots, and conservation. *Proceedings of the National Academy of Sciences*, 103(51), 19374-19379.

Chen-Kraus, C., Raharinoro, N.A., Lawler, R.R. and Richard, A.F. (2023) Terrestrial tree hugging in a primarily arboreal lemur (*Propithecus verreauxi*): a cool way to deal with heat? *International Journal of Primatology*, 44(1), 178-191.

Clutton-Brock, T.H. and Albon, S.D. (1979) The roaring of red deer and the evolution of honest advertisement. *Behaviour*, 69(3-4), 145-170.

Cooke, R.S., Eigenbrod, F. and Bates, A.E. (2019) Projected losses of global mammal and bird ecological strategies. *Nature Communications*, 10(1), 2279.

Donlan, C.J., Berger, J., Bock, C.E., Bock, J.H., Burney, D.A., Estes, J.A., Foreman, D., Martin, P.S., Roemer, G.W., Smith, F.A. and Soulé, M.E. (2006) Pleistocene rewilding: an optimistic agenda for twenty-first century conservation. *The American Naturalist*, 168(5), 660-681.

Eisenberg J. F. (1981) *The Mammalian Radiations*. University of Chicago Press, Chicago, IL.

Elemans, C.P., Jiang, W., Jensen, M.H., Pichler, H., Mussman, B.R., Nattestad, J., Wahlberg, M., Zheng, X., Xue, Q. and Fitch, W.T. (2024) Evolutionary novelties underlie sound production in baleen whales. *Nature*, 627, 123-129.

Farnsworth, A., Lo, Y.E., Valdes, P.J., Buzan, J.R., Mills, B.J., Merdith, A.S., Scotese, C.R. and Wakeford, H.R. (2023) Climate extremes likely to drive land mammal extinction during next supercontinent assembly. *Nature Geoscience*, 16(10), 901-908.

Food and Agricultural Organisation (2020) *The State of the World's Forests. Forests, Biodiversity and People*. FAO, United Nations.

Gómez, J.M., Gónzalez-Megías, A. and Verdú, M. (2023) The evolution of same-sex sexual behaviour in mammals. *Nature Communications*, 14(1), p.5719.

Greenspoon, L., Krieger, E., Sender, R., Rosenberg, Y., Bar-On, Y.M., Moran, U., Antman, T., Meiri, S., Roll, U., Noor, E. and Milo, R. (2023) The global biomass of wild mammals. *Proceedings of the National Academy of Sciences*, 120(10), p.e2204892120.

International Rhino Foundation (2023) 2023 State of the Rhino Report. Available at: https://drive.google.com/drive/folders/1B_K0n3YTtdaU6I08dNCq3H0JP45jvfCO (accessed 30 May 2024).

Jakes, A.F., Jones, P.F., Paige, L.C., Seidler, R.G. and Huijser, M.P. (2018) A fence runs through it: A call for greater attention to the influence of fences on wildlife and ecosystems. *Biological Conservation*, 227, 310-318.

Jepson, P., Deaville, R., Barber, J. *et al.* (2016) PCB pollution continues to impact populations of orcas and other dolphins in European waters. *Scientific Reports* 6, 18573 (2016). DOI: 10.1038/srep18573.

Keast, A. (1969) Comparisons of the contemporary mammalian faunas of the Southern Continents. *The Quarterly Review of Biology*, 44, 121-167.

Kuhar, C.W., Fuller, G.A. and Dennis, P.M. (2013) A survey of diabetes prevalence in zoo-housed primates. *Zoo Biology*, 32(1), 63-69.

Laumer, I.B., Rahman, A., Rahmaeti, T., Azhari, U., Hermansyah, Atmoko, S.S.U. and Schuppli, C. (2024) Active self-treatment of a facial wound with a biologically active plant by a male Sumatran orangutan. *Scientific Reports*, 14(1), p.8932.

Lehman, S.M. and Fleagle, J.G. (2006) Biogeography and primates: A review. In: Lehman, S.M. and Fleagle, J.G. (eds) *Primate Biogeography*, Springer, New York, pp. 1-58.

Macdonald, D.W. (2019) Mammal conservation: Old problems, new perspectives, transdisciplinarity, and the coming of age of conservation geopolitics. *Annual Review of Environment and Resources*, 44, 61-88.

Malhi, Y., Lander, T., le Roux, E., Stevens, N., Macias-Fauria, M., Wedding, L., Girardin, C., Kristensen, J.Å., Sandom, C.J., Evans, T.D. and Svenning, J.C. (2022) The role of large wild animals in climate change mitigation and adaptation. *Current Biology*, 32(4), pp.R181-R196.

Mammal Society (2024) Available at: https://www.mammal.org.uk/species-hub/uk-mammal-list/ (accessed 30 May 2024).

Manca, C., Boubertakh, B., Leblanc, N., Deschênes, T., Lacroix, S., Martin, C., Houde, A., Veilleux, A., Flamand, N., Muccioli, G.G., Raymond, F. *et al.* (2020) Germ-free mice exhibit profound gut microbiota-dependent alterations of intestinal endocannabinoidome signaling. *Journal of Lipid Research*, 61(1), 70-85.

Moleón, M., Sanchez-Zapata, J.A., Donazar, J.A., Revilla, E., Martin-Lopez, B., Gutierrez-Canovas, C., Getz, W.M., Morales-Reyes, Z., Campos-Arceiz, A., Crowder, L.B. and Galetti, M. (2020) Rethinking megafauna. *Proceedings of the Royal Society B*, 287(1922), p.20192643.

Nicholson, P. T., Ikram, S. and Mills, S. (2015) The catacombs of Anubis at North Saqqara. *Antiquity* 89 (345), 645-661.

Packer, C. (1977) Reciprocal altruism in *Papio anubis*. *Nature* 265(5593), 441-443.

Packer, C. and Pusey, A.E. (1982) Cooperation and competition within coalitions of male lions: Kin selection or game theory? *Nature*, 296(5859), 740-742.

Payne, K.B., Langbauer, W.R. and Thomas, E.M. (1986) Infrasonic calls of the Asian elephant (*Elephas maximus*). *Behavioral Ecology and Sociobiology*, 18, 297-301.

Rauber, R. and Manser, M.B. (2018) Experience of the signaller explains the use of social versus personal information in the context of sentinel behaviour in meerkats. *Scientific Reports*, 8(1), p.11506.

Remport, L., Sós-Koroknai, V., Hoitsy, M. and Sós, E. (2022) Mandibular fractures in giraffes (*Giraffa camelopardalis*) in European zoos. *Journal of Zoo and Wildlife Medicine*, 53(2), 448-454.

Ritzel, K. and Gallo, T. (2020) Behavior change in urban mammals: A systematic review. *Frontiers in Ecology and Evolution*, 8, p.576665.

Schmidly, D.J. and Naples, V. (2019) North American mammalogy: early history, dominant personalities, and significant milestones (1850–1960). *Journal of Mammalogy* 100(3), 701-718.

Sinclair, A.R.E., Fryxell, J.M. and Caughley, G. (2006) (2nd ed.) *Wildlife Ecology, Conservation, and Management*. Blackwell Publishing, Oxford.

Skogland, T. (1985) The effects of density-dependent resource limitations on the demography of wild reindeer. *The Journal of Animal Ecology*, 54, 359-374.

Understanding Animal Research (2022) *Animal research statistics for Great Britain, 2022*. Available at: https://www.understandinganimalresearch.org.uk/news/animal-research-statistics-for-great-britain-2022#:~:text=The%20figures%20show%20that%202%2C761%2C204,of%20all%20procedures%20in%202022 (accessed 14 May 2024).

Vynne, C., Gosling, J., Maney, C., Dinerstein, E., Lee, A.T., Burgess, N.D., Fernández, N., Fernando, S., Jhala, H., Jhala, Y. and Noss, R.F. *et al.* (2022) An ecoregion-based approach to restoring the world's intact large mammal assemblages. *Ecography*, 2022(4).

Wallace, A.R. (1852) On the monkeys of the Amazon. *Proceedings of the Zoological Society of London* 20, 107–110.

Wenker, C.J., Stich, H., Müller, M. and Lussi, A. (1999) A retrospective study of dental conditions of captive brown bears (*Ursus arctos* spp.) compared with free-ranging Alaskan grizzlies (*Ursus arctos horribilis*). *Journal of Zoo and Wildlife Medicine*, pp.208-221.

Williams, C.T., Barnes, B.M. and Buck, C.L. (2016) Integrating physiology, behavior, and energetics: biologging in a free-living arctic hibernator. *Comparative Biochemistry and Physiology Part A: Molecular & Integrative Physiology*, 202, 53-62.

Wilting, A., Courtiol, A., Christiansen, P., Niedballa, J., Scharf, A.K., Orlando, L., Balkenhol, N., Hofer, H., Kramer-Schadt, S., Fickel, J., Kitchener, A.C. (2015) Planning tiger recovery: Understanding intraspecific variation for effective conservation. *Science Advances*, 26 Jun 2015: Vol. 1, no. 5, e1400175. DOI: 10.1126/sciadv.1400175.

Printed and bound by CPI Group (UK) Ltd, Croydon, CR0 4YY

18/12/2024

14614318-0001